# the trinity theory

# the trinity theory

## Vol. I
### The Human Science of Soul

## Trinity Sarah Craig

First published in 2015 by Trinity Artistix Corporation
Victoria, British Columbia
www.trixxcorp.com



Library and Archives Canada Cataloguing in Publication


Craig, Trinity Sarah, 1974-, author
    The trinity theory / Trinity Sarah Craig.

Includes bibliographical references and index.
Contents:  vol. 1. The human science of soul
Issued in print and electronic formats.
ISBN 978-0-9940018-0-1 (v. 1 : pbk.).-- ISBN 978-0-9940018-1-8 (v. 1 : pdf)


    1. Spirituality.  I. Title.


BL624.C73 2014                    204                    C2014-908155-3

                                                       C2014-908156-1

••• re mem ber •••

*The Trinity Theory* explores the human science of soul. It presents a unique point of view, which demonstrates how you can strengthen your ability to create yourself throughout your lifetime. By showing you how to focus on having a passionate connection to life, it offers you ways to bring balance between both your dark and light qualities. It brings new twists to old concepts by revealing fresh perspectives on ancient ways. It unearths spiritual secrets that have not been available until this jeweled key has unlocked them.

*The Trinity Theory* is an energetic guide to understanding life on Earth. It is composed of 13 concepts that are distinct, yet harmonic aspects of the cosmology of life. They present genuine guidelines for understanding your place in the Universe. The core principles within flow from the Laws of Nature and represent the inherent formulae of the Creatrix that govern your life. The Theory is simple to understand, yet deep in meaning.

*The Trinity Theory* is bold. It is unlike any other spiritual system available. With a foundation of the spiral, it is an expression of the instinctive, cursive nature of all living beings. It gives ways to connect with the spiral in your life so you may live in harmony with the world around you, empower your ability to have a strong soul connection and make it easier to find personal freedom. It relates a complete way of life by touching upon many facets of your journey, exploring ways to improve the quality of your experience and revealing how to hold your head so you can see the rainbow.

# ✦✦✦ table of contents ✦✦✦

## six ◆◆◆ alchemy . . . . . . . . . . . . . . . . . . . . . . . . . . . 57



## seven ◆◆◆ rainbow . . . . . . . . . . . . . . . . . . . . . . . . . 95

## thirteen ◆◆◆ mystery .................................. 201

# ✦✦✦ introduction ✦✦✦

Over time, in the stagnant struggle of duality with its Dark/Light, Good/Bad and Devil/God, humanity has lost something precious. This hidden gem, the Trinity Theory, is an ancient perspective, a third way of being and another option for living based upon the primordial Laws of Nature and their sublime foundation, the spiral.

Straight lines do not exist in Nature. Duality, with its linear, human-made concepts uses cubic systems to manage a naturally curvaceous and spiralized world, as every living being in the Universe grows according to the same blueprint of a spiral. This contradiction is the heart of the conundrum humanity faces today, made obvious by the turmoil observable in many aspects of society.

*The Trinity Theory* is a way to perceive the world that touches upon many vital facets of human existence. It is a theory that teaches you to acknowledge the spiral in your reality in order to work around the inflexible nature of the square systems that pervade reality. It will allow harmony to flow into your lifestyle, enliven your world-view, and improve the quality of your life.

*The Trinity Theory* offers a way of being. It is a spiritual current that demonstrates one way to climb the ladder of divinity. It is comprised of 13 chapters that begin simply, and grow outwards, upwards, and inwards, somehow finding their way back to the beginning again. It weaves a web of perceptions to be enjoyed, a wonderful cave full of gold, jewels, and gems for you to hunt.

*The Trinity Theory* is a system of being based upon the idea that every living thing in the Universe is a spiral. Humans and all the creatures of the Earth are spirals, the official blueprint of Nature. Solar systems, cells, your fingertips and your cat are all living, functioning spirals. You know the spiral, you live the spiral, and you are the spiral.

Society with its linear systems is not based upon the spiral. In fact, it runs counter to the spiral because it is based upon the lifeless concept of duality. Duality is straight, polarized, and flat. This has created much discomfort and unrest for you because it dictates a lifestyle that is far removed from your natural way of being.

The pantheon of love/hate, wrong/right and bad/good have held you under their spell for millennia now. *The Trinity Theory* is here to banish that illusion and replace it with old-fashioned truth.

*The Trinity Theory* synthesizes many separate forms of spirituality into a functioning whole. In the past, these systems have been practiced separately, and the divisions can be seen today amongst people. Division can only ever create weakness and conflict, as is evident in the state of the world today.

*The Trinity Theory* knows that soul is soul, and there are no divisions within it. Real power comes from acceptance. This is a time when the spirit people of the world can unite, so that the power of knowledge can become available to all.

You are an Angel. In other words, you are an immortal human being. You may not know it. You may not remember it. But, it is true. Humans comprise one of the oldest, purest and most powerful bloodlines in the Universe. *The Trinity Theory* is an energetic guide to remembering your Angelhood. It will show you how to unfold your wings.

Welcome to The Trinity Theory.

## ◆◆◆ greetings ◆◆◆

Greetings.  My name is Sarah Jane Craig, which means gracious princess of the rocks, which is clever because I really love rocks.  My friends call me Trinity.  I was born on August 30th, 1974 in Wimbledon, England.

I come from a Canadian military family, stationed in England when I was born.  I am mostly Scottish in heritage.  My great-grandfather, Sir James Henry Craig, five times removed, was the Governor-General of the Canadas and Lieutenant-Governor of Lower Canada from 1807 to 1811.

By the time I was 17, I had lived in four countries: England, Canada, the United States and Germany, travelled in 17 countries and had gone to eight schools.  When I was in grade two, I pulled a straight pin out of a light socket on a dare.  I was blown across the circle at story time.  It caught the wall on fire and made all the kids cry.  I think that lightning bolt was the beginning of my belief in a higher power.

I have included many things my life has taught me in the spiritual system that constitutes this book.  I have given what I have and what I am to these pages.  I want those who come after me to have a record of my spiritual work to inspire them to create their own.  Maybe they will share it too.  By sharing spirit in this way we make it grow.  Soul is contagious.

I am an artist and a self-published author.  I wrote *The Trinity Theory* and created all of the artwork, formatted the 3 different formats, published the book, made the trailer video, narrated the trailer video, made the music for the trailer video and designed my website.

One is you.  We are One.  You have a heart.

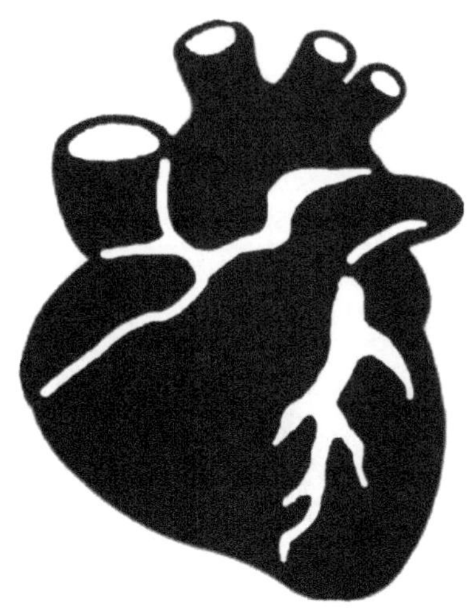

# the Human Heart

Being human is the ultimate experience.  As a human, you have been given the most powerful gift in the entire Universe, the omnipotent tool that is heart.  Your heart is your ancient and sacred source of primordial energy, a volatile core of power within, the eye of your storm.  It is the continual fount of divine universal energy, which creates the sparkle in your eye.  The heart is a gateway to the soul.

You have a personal fountain of youth that guides you as a lantern through the night, parting the mists.  It reveals the perfect path for you to take.  Your heart is the brightest, strongest and most accurate guiding force in your life.  It is always connected to the sacred mystery, the never-ending source of life.  It freely offers up its divinity to you.

Your heart never lies.  It never cheats you.  It accepts you, no matter what you do.  It is the enigma of you that never fails.  A part of you that is right here, always available, forever by your side, even when you do not want it to be.  It has the ability to forgive, to make the best choice, and to carry on faithfully.  Your heart is an emotional organ.  It beats faster when you are excited or afraid.

The heart works with the unconscious mind, therefore has incredible depth.  From there it creates your consciousness, your personal patterns, and your attitude.  Your heart is a bit of a dreamer.  But, that is okay, let it dream.  These dreams are yours for the taking.  They are exactly what you really want.  It is wise not to fear these primeval longings.  They are only your innermost workings sharing themselves with you.

Truth is important to the heart.  It naturally gravitates towards things that are true for it.  You have an ability to feel your truth because it makes you tingle.  Your heart is concerned with truth because that is the energy most closely resembling it.  It resonates with authenticity.  The energy of your truth rings like a bell, to be heard by those who are listening.

As the beginning point of all health and all relationships, the heart is where healing happens.  When your heart has made up its mind, that is that, there is no way to dissuade it.  There is no point in trying to argue with your heart, it does not have ears to hear.  It is the centre where change is born.

You can use your heart as a tool of communication, by speaking from the heart.  When you come from the heart, it assures you of earnest, direct and real transmissions.  At the other end, your heart will always react to the validity of what it receives from others.  Your heart has a built-in lie detector.  If you listen carefully, it will tell you how dependable the character in front of you is.

Your heart is like a compass.  It points to the exact path between all the paradoxes of life, the things you spend your time wondering: Should I do this, or that?  When you have the daring to follow your heart, trusting it instead of doing your duty, everything becomes easy.  You develop a natural confidence that enables you to deal with life gracefully.  When you are in service to your heart, things have a funny way of working out.  This is what it means to have an open heart.

## energy

Your heart is intimately connected with energy.  Energy is the language of the heart.  Your heart uses energy to communicate the way it feels, as a painter does with colour.  Your heart expresses itself through the energy of feelings.  Energy conveys both your inner realms and outer realms, with its tangible and perceivable presence.

Subtle energy creates an etheric atmosphere of universal life force. It is present in, around, and between all things. Flowing, constantly shifting and changing, like clouds in the sky. Energy is everywhere.

You know energy is real because you can see with your eyes how the dowsing rod will dip when there is an indicator of ground water. You can then dig to find the physical evidence. Animals use the energy of the magnetic lines of the Earth when they migrate, for how else would they be able to navigate such great distances? There have been medicinal practices concerned with energy for thousands of years.

Try pushing two polar magnets together. You can feel the magnetic energy as it repels them. What about lightning? A perfect example of the power of energy and its ability to create something out of nothing. You can experience energy yourself with a pendulum. Watch as with a bit of practice you can control its direction and swing. These examples tell you that there are energetic forces you cannot see at work all the time all around you.

Energy is observable like mist. Most people cannot really see it, but you can certainly feel it. You can perceive it with your senses by tuning into it. It is not a rational thing. You cannot decipher it with your mind. Instead, you must feel it in order to understand it.

Energy is how an environment feels when you walk into it. It is the tangible tingle that you can feel when you encounter another. Energy is how you feel, how that person feels to you, but also how the atmosphere, the colours, the furniture and the air quality makes you feel. It is many feelings that are mixed together, offering themselves to you when you enter into a situation.

For example, you walk in a room and you feel a palpable energy in the room like an electric charge. There is a movie star in the room and you are feeling the energy of their presence as the excitement of the crowd is crackling in the air. Or when you move into a new house. You can still feel the residual vibrations from the people who used to live there. This kind of pervasive energetic leftover is present all the time. You just have to tune into it by bringing your awareness to it.

Reading energy is discerning how something makes you feel when you come across it. It is not you, but the situation that can cause feelings in you. Understanding your own feelings and reactions is a method of discerning the energy of different things. Energy is tangible, yet subtle. Experiencing it is different for everyone. People all have different ways of interpreting energy.  Some do not feel it at all.  Some feel it so much that it is difficult for them to get out of bed.  Everyone leaves a rainbow slime trail of energy wherever they have been, for anyone to see.

When you meet someone new, you put out your feelers and feel them, to know if you like them or not.  Everybody does it.  Everybody can do it.  It is just a matter of being conscious of it or not.  With practice, you can use your antenna in all situations, to discern the energy behind them.  This can be beneficial because it can give you warnings and indications of things to come.

Reading energy is kind of like going to an opera sung in Italian.  You may have no clue what they are saying, but you still get the idea.  You can feel the emotions and read what is happening between the performers, though you may not consciously know what is being said.  Communication with others is much more than mere words.

Energy can come from people, physical contact, the physical environment, living beings, from objects, intense emotions, or it can come from the past, as a few examples.  Energy has layers upon layers, each changing the other.  It accumulates over time.  Energies come together and mingle, creating new expressions of energy.  It is a constant and unending upheaval.

When going for a walk, you may see some rocks up ahead that do not seem quite right.  Oh wait, those are not rocks.  They are horse apples.  You can definitely feel the energy difference between the two, as fresh animal droppings vibrate intensely.

Everything has its own unique energy signature, its own vibration.  Something vibrates quickly, slowly or somewhere in between.  Things tend to resonate with each other or they repel each other.  That is why you can feel things click when you come across something that works for you.

Inanimate objects can take on the energy of time and hold energetic impressions whether wood, fabric, gold, plastic, steel or glass.  Objects can hold impressions of feelings and emotions, which a sensitive person can read via psychometry.

Causality, a Law of Nature, says that every cause has an effect.  So you might ask yourself, what is the cause of love?  It must have its origins in energy.  How do you know when you have to go pee?  You get a feeling.  Your feelings are always communicating to you about the energy of your environment. All you have to do is be sensitive to them.  Just listen, they are speaking to you.

## aura

Your heart is a luminary that radiates light, shining like a star in the night sky. This light is your aura. It is an extension and an energetic expression of your heart. You are constantly emitting colours, which reflect your state of being, including your feelings, beliefs, and thoughts.

Your feelings have an energetic counterpart in the etheric realms of life where they are quite alive, visible as colours. Your aura looks like a shimmery rainbow egg, flowing with colours that constantly shift as your feelings change.

Your aura is made of your radiant life force that defines your personal space. It is your protective defense against the energies of the outside world. It represents the boundary between your inner and outer realms. Your aura is a strong castle around you. It is your personal golden egg of protection, as it will not let outside energies in, unless you want them to enter.

Your aura communicates and displays the energy of the way you feel, much like the weather. When you are clear with your feelings, then your aura can be luminous. When you are heavy with feelings, your aura can be dark like storm clouds. Think of how quickly your feelings change, and this is how quickly the aura flashes and fades.

Your aura is a colour filter wrapped around you. Like a hued lens, it colours your vision. It is a tinted glass and it distorts the way you see things.[1] Your beliefs, feelings and thoughts shade everything you perceive in your reality. They give you biased feedback. The green you are looking at turns bluish as you peer through the yellow of your aura. In this way, you know that your beliefs affect your reality.

The aura is the body's electromagnetic energetic system. It taps into the boundless, natural electromagnetic energy of the Universe.  A subtle bioenergetic system is capable of perceiving the vibrations of other subtle energetic systems.  Your aura is your intermediary with all things energetic. Everything living has an aura.  Science is even able to take pictures of it.

The aura can be seen, but felt even more.  When you are paying close attention there can be a tangible atmospheric change when you walk into the boundaries of another's aura.  It is inevitable that auras intermingle with each other.  It is through your aura that you are able to feel the energy of other people.  As auras brush together, you can sense the moods of others.

You can see the auras of trees, plants, friends and animals if you really look.  It helps to use a light background, or a dark one.  It is a simple matter of relaxing the eyes and being patient.  Let your eyes go unfocused. You can see it, but you have to allow yourself to see it.  It is like when you look at something bright and there is an afterglow in your vision.  The aura of the tree appears like a coloured after-burn that frames the tree.

Your aura displays the energy of your wisdom, knowing, feelings, unconscious, Will, spiritual connection, intensity, thoughts, desires, dreams, visions, responsibilities, health, experiences, and spiritual knowledge.  You are but a drop of water in a storm of energy.

## the energetic world

The energetic world is a mirror world that lives and breathes, cheek to cheek, with this, the waking world.  Just as your conscious mind works in tandem with your unconscious mind, so too does the real world operate in harmony with the spirit realm.  Just as iron filings dance along the magnetic lines of a magnet, so too does your energy affect physical reality.

The Law of Conservation of Energy, a Law of Nature, says that energy never dies, it just changes form. It says that in a closed system (like the Universe) that the total amount of energy stays constant. Energy is clearly present in your life. It is a reflection of your physical world.

The energetic world is known as the realm of the imagination, the dream world, the flipside, the astral realm, the collective unconscious, the Underworld, the Otherworld, the Netherworld or the Summer Country. The energetic world is backwards and upsidedown from you, kind of like the internal mechanisms of a camera.

All life in the Universe is connected and affected by this energetic ocean. All the thoughts, dreams, beliefs, visions and daydreams of all the people who have ever lived are alive in the astral realm. It is like a movie that does not stop, comprised of your imagination telling a magnificent story of your passions, desires, doubts and fears. Your unlimited creative potential is combined with your greatest nightmares, forming a rich, thick cosmic soup.

According to the field of mathematical study called the chaos theory, there is a concept known as the butterfly effect. It is a theory that says the fluttering of the wings of a butterfly can cause a tsunami on the other side of the world. What you do affects every other living being. What they do, affects you.

The energetic world also spans all of time. The past, the present and the future are all one, and each affects the other. In this energetic domain, humans are all One because our energies are so tightly intertwined. Nothing is ever forgotten. The energy is cumulative over time in the spirit realm, like the pages of an old book awaiting to be read.

The energetic world reflects back exactly what is going on. Energy never lies. It reveals the truth of the waking world, back to itself. No matter what the real world says, it is wise to trust what the energy is telling you first. The energetic world is where ideas are born, where thought forms take on life and where the colours of your aura dance and play. The energetic world is where you go when you are dreaming.

It is wise to accept the fact that every thought you have is a creature that is alive and waiting to be fed. Some humans can see your thoughts as easily as you read these words. All of your thoughts and beliefs gather around you in the energetic world. Like a endless story, they wrap around you, vibrantly colourful and conscious. You are like a bigtime director in the energetic realm, retelling the tale of your life with every nuance and expression.

The etheric realm is where your imagination lives. When you imagine things, they become real. They grow and expand, especially when you feed them the energy of your thoughts and attention. It is wise to take responsibility for the realm of your imagination because it affects others.

Everything is seeded and begun in the energetic world. From there it progresses to manifest physically in the material world. Everything began as a mere seed in the spirit world before birthing into the waking world. You know this because all matter is really condensed energy. The Theory of Relativity, a Law of Nature, states that energy and mass are two names for the same thing.

The spirit world is an ocean that is chaotic, unpredictable and infinite. All of your thoughts, dreams and fantasies go into this world where they swirl around, waiting to be born or to die. It hosts many illusions, truths, and mirages. This world is full of half-formed ideas that can be developed if you truly desire it. It is a realm of pure energy where all beliefs are tested out to see if they stand the test of time.

The energetic world will throw your truth back to you. It reveals the paradoxes and discrepancies of the world around you. It reflects back these things by subtly triggering your senses. Your friend might be saying one thing, but their energy is giving you the nagging sense they really mean something else. Your senses are your antenna. You can use them to perceive the energy of what is truly going on in your reality, to find the deeper meaning.

It is wise to marry yourself to the energetic world by becoming one with your senses and your energy. You can learn to flow with your instincts, listening to their ancient wisdom. The more sensitive to your senses you become, the easier it is to follow their flow, heeding their guidance. What it seems like, what it feels like, is exactly the way it is. No matter what your brain is trying to tell you.

Through your senses, you can connect with the energetic realm. This intense world wants to empower you with what it knows. It wants you to be brave enough to pick up your intuitive tools and use them as a flashlight to shine into the depths of your own ignorance. Once you have harnessed their power, they can make you a force to be reckoned with.

Energy, like water in the ocean, flows in currents. These currents of energy flow as cultural beliefs, pools of thought, and ways of being. Many humans tend to share the same feelings and have done so throughout time. These currents run deep and you participate in them without conscious thought.

The energetic world is full of elemental forces. They are profound, primordial energies with ancient swirlings and seductive tides. They ebb and flow with the powers of destruction and regeneration. This ocean of energy has an intensely penetrating effect upon your life. Awareness to energy is a key to life.

One represents the integrity that you have with yourself.

Two is balance within and without.  You have a pair of eyes, ears, hands, legs, lungs, arms, feet and two hemispheres of the brain.

## Goddess and God

Goddess ◆ God

Dark ◆ Light

Matter ◆ Spirit

Mother ◆ Father

Magnetic ◆ Electric

Receptive ◆ Dynamic

Negative ◆ Positive

Black ◆ White

Minus ◆ Plus

Fusion ◆ Fission

Heavy ◆ Buoyant

Opaque ◆ Clear

Dense ◆ Loose

Widdershins ◆ Sunwise

Night ◆ Day

Woman ◆ Man

Mass ◆ Energy

Convection ◆ Conduction

Earth ◆ Sun

Particle ◆ Force

Electron ♦ Proton

Attraction ♦ Repulsion

Alkaline ♦ Acidic

Black crystals ♦ White crystals

Form ♦ Force

Space ♦ ime

Implosion ♦ Explosion

Centrifugal ♦ Centripetal

Inertia ♦ Desire

In ♦ Out

Goddess is the Dark.  She is everything material.  Her domain is all the planets, stars, elements, animals, plants and particles of the Universe. For humans the Goddess is planet Earth.  Her spirit resides in the iron-core crystal at the centre of the Earth.

God is the Light.  He is the electric spirit that moves through matter. He is the current that begets a heartbeat.  For humans God is invisible waves.  His spirit resides at the centre of the Universe.

## the Creatrix

Together Goddess and God are the Creatrix, Prime Source, the Great Spirit, Source, the Universe, and Prime Mover.  Together they create soul, colour, heat, energy, beauty, lightning, and life.  They are the source of the sparkle. They are androgynous, a hermaphrodite and a composite being.

There is only one way, and that is Sacred Law of the Creatrix.  Sacred

Law is expressed upon Earth as the Laws of Nature. The Creatrix has given us the Laws of Nature as a subtle, yet effective guide to life. These Laws are the foundation for life on Earth. They are perfect. They are right. They are the truth. They were not made by humans. Therefore, they can be trusted to have the benefit of the whole as their prime intent.

They are One. They are in everything. Everything is in them. They complete each other. They are not separate. You cannot have one without the other. One is neither greater, nor lesser than the other is.

Goddess and God are equal yet opposite expressions of the same thing, vibrating at different speeds. They represent a Creatrix Circuit. Matter gives the illusion of solidity but it is a highly vibrating condensed energy itself. The current of spirit is invisible, but without it, there would be no life.

Another example of a Creatrix Circuit is the Earth and the Sun. The Earth pulls with her gravity from her iron-core crystal, the Sun sends with his radiation from his nuclear core. As a result, life flourishes.

The Creatrix is within you. You are a bioelectromagnetic being filled with electromagnetic energy. Both electricity (God) and magnetism (Goddess) are the foundational energies present in your body that make it function. Your brain has two opposite hemispheres, a female half and a male half, working in tandem. Another example of this is the union between a woman and a man, the result being a baby.

A magnet is an example of a Creatrix Circuit, with its North and South poles. There are positively charged substances, and negatively charged substances. Voltage can have either energy or the loss of energy. There are magnetic fields or electric fields. Ions can be negatively, or positively, charged. Molecules have negative and positive poles. You find the Creatrix Circuit throughout Nature. Cells, atoms, the human body, the Earth and the Sun all have a North and South pole.

Goddess and God are not physical floaty dudes in the sky, or sitting on thrones under the ground. They are scientific principles that represent the energies of ebb and flow. They are something that is commonplace throughout the Universe. It is present with you in even the simplest of tasks. They are not flakey, hokey or even spiritual. They are natural and familiar forces you deal with every day.

Everything the Creatrix exists in is alive. For example, light is alive. This is known because it has the ability to change from a wave to particles by its own will. Darkness is also alive. This is known because of how it makes you feel as you race home after a scary movie. Sound is alive. You know this by its ability to be in many places at once. It can be difficult to pin down its source.

You are a tiny cell in the huge body of the Creatrix. You are an individual aspect that together with the rest of universal life collectively makes a whole. Everything is perfect because everything is the Creatrix. It is neutral. It is limitless. It is absolute.

The Creatrix is an awesome thing. Looking at the perfection of the world around you. It is difficult not to be overwhelmed at the beauty and symmetry of the Creatrix's garden. Every day science is humbled by the power of Nature. The Creatrix is a subtle, yet strong leader and a fantastic role model. The Creatrix is an adept ringleader of this incredibly huge and complex system called the Universe.

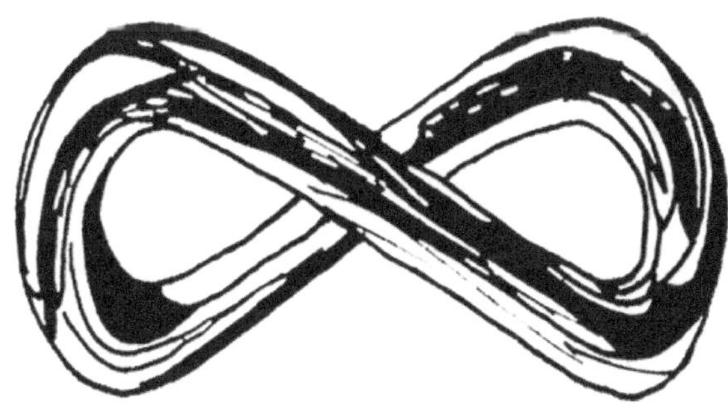

## Heartbeat

The Creatrix has a heartbeat that is the pulse of the Universe. The heartbeat is created by the exchange of energy between Goddess and God. She pulls and he pushes. This sway of energy occurs within every living being, molecule, breath, body, force, everything.

The energy flows like an infinity symbol, from one-half to the other, as inhalation follows exhalation. The energetic exchange created by the heartbeat is called ratcheting. One side is empowered after the other and back again.

When the energy of the female and male are in balance, together they create a Creatrix Circuit. One energy causes the other to happen. Meaning they cannot function without each other. They are a self-propelling energy circuit.

Consider electromagnetic energy. It is a force made of two forces that harmoniously work together. It is the foundation for all life. He is electric. He pushes. She is magnetic. She pulls. This ratcheting of energy from one pole to the next is the heartbeat of the Creatrix. Things become magnetic with magnetism. Then they are able to hold charge or electricity. Then things become electric and are able to hold a magnetic field. Then it switches. This is the founding principle of life.

Ratcheting is the foundation of growth in the Universe. An undulating energetic current is created as the energy travels between the female and the male, growing and amplifying with each turn. This pulse generates every cycle, from cellular development, to love, to how flowers grow. This is the primal ebb and flow of the energy within the Universe. It creates all life.

Life functions with the energy of turmoil turning to relaxation, tension to slack, restriction to limpness. A wise person knows to live their life following this pattern. Variation is a foundation of vital living. Fluctuating between things as a lifestyle is following the Laws of Nature. The undulation of the energetic current of ratcheting creates momentum in your life that can be used to propel you.

## grounded

To be grounded means you are fully present in your body. It means you are connected to the Earth and the Goddess through your choices and actions. She is magnetic, and being grounded is your ability to have and hold magnetism.

Just as lightning or electricity grounds in the Earth, you need that connection underneath so you can function properly. You are like a storm in that you are an electrical being that needs to ground your abundant energy.

Humans ground from the root, not from the feet. This means that you have energetic cords like roots that extend from your perineum, the place right between your legs, far into the ground. The more grounded you are, the deeper the roots go. They grow with your awareness and attention. These roots are one way that you can communicate with the Earth and therefore the Goddess.

When you are grounded, you are in a state of equilibrium. You are prepared, calm, well fueled, empowered, totally present, accepting, and present in the now. You are not easily distracted. It means you are in control of your energy and are not dependent upon outside sources to refuel.

Grounding makes you alert, focused and aware of yourself and your surroundings. It enables you to discern the truth about people around you. It aids in not becoming overwhelmed. It heightens your sensitivity. It assists you in trusting your feelings more easily, and it empowers your natural abilities. Grounding offers you the ability to connect to the Earth, to maintain and fuel your vibrancy.

When you are grounded, it leads to acquiring common sense. This is a valuable tool for your life because it enables the highest productivity in accomplishing your desires. Being grounded in the Earth gives you the ability to fly high, just as the root system of a large tree balances the tree, enabling it to grow tall in the sky.

To be ungrounded means that your Emotional Body is not in your Physical Body, but instead hovers outside of it. It is easy for a sensitive person to see eyes floating above the head of an ungrounded person. To be ungrounded means you can be flighty, spaced-out, forgetful, not hungry, not present, easily distracted, jittery, uncaring, you are prone to accidents or dropping things, you arrive late, have a difficult time concentrating, and tend to live in a fantasy world.

You can ground yourself by doing a grounding ritual, connecting with the Earth, clapping your hands, stomping your feet, hugging a tree, petting an animal, touching another human, experiencing pain, holding a magnet, eating some food, stretching your body, breathing deeply, touching the ground, removing your shoes, using memories of Nature, going for a walk, gardening, cleaning something, going away from technology, going outside, having a foot bath, making something with your hands, cooking, or carrying rocks in your pocket.

You can also become too grounded and inertia can set in. This causes you a difficult time moving and getting on with your life. This unbalance can be a quick path to an unhealthy lifestyle.

Some people are naturally more grounded than others. It is wise to make being grounded a priority in life. It cannot be over-stated as to the importance of actually being in your body and connected to the Earth below you. All of life stems from its relationship with the Earth. Without the Earth's support you leave yourself unprotected in a harsh world. Make sure she has your back.

So many problems the world today faces is because so many humans have lost their connection to the Earth. If we were more connected we would be less likely to tolerate the pain and suffering that the Earth and her creatures are experiencing.

# enlightened

Just as the Earth receives energy from the Sun, so do you receive energy from God, the centre of the Universe and the Sun, in the form of universal life force or prana. He is dynamic, and enlightenment is your ability to have and to hold electricity, or charge.

Prana descends energetically through the top of the head, flooding the body with the essence of life. Without prana, you would die. Prana is needed by all parts of the body for proper functioning.

Being enlightened with this universal energy increases your access to light, which is knowledge. This light of information is the lightning of illumination that stimulates ideas, innovation and learning. This energy is lofty energy that can assist your aspirations, aid in rising above the refuse, and bring clarity. Prana is inspiring, uplifting, and freeing. By having a connection with the centre of the Universe, you are tapping into the limitless information that the Universe freely offers up. There is much to learn on the God waves.

Sunshine is fuel. Illumination is enlightenment. You can brighten your world and this clarity can show you the connection between things. You become able to see the interconnectedness between all life. This is traditionally called wisdom.

Becoming enlightened is not as simple as becoming grounded. Some ways to bring more prana into your crown are reading, meditating, being free from technology, exercising, experiencing Nature, eating whole foods, dancing, drinking water, and doing breathing exercises, to name just a few. You can also acquire too much light. Then you can become a burnt out or flakey person, causing you to have a difficult time concentrating and understanding things.

# sides of the body

The marrying of Goddess and God is present in the body of humans as you have two halves to your brain. The brain has a female half and a male half. They are both equal, yet opposite expressions of the same thing. They are different, yet both equally important to the functioning of a healthy human. Like the poles of a magnet, they work together in harmony.

It is wise to create bridges between the two halves of your brain so you can use both equally. This balance of the hemispheres working together in tandem creates harmony in life. A fully active brain results in enhanced cognitive functions.

A great way to bring balance to the dual sides of the brain and body is to practice using both hands in various tasks like playing Frisbee, writing, using your mouse, and brushing your teeth. It is wise to try to do things unhabitually, or in ways you do not normally do them. You can enjoy the challenge.

This will make you use parts of yourself that are not usually active in that way. By giving the other side of your body a chance to express itself, you are using way more muscles than normal. This forces the brain to fire in unusual ways, directly expanding it, and making it more open. A tricky way to get more clever.

When the two halves of your brain are balanced, there is no obstruction to slow things down so your life flows easily. You can learn to be whole by integrating both halves of your brain. You are androgynous, just like the Creatrix.

## female

The right hemisphere of the brain is female. It is creative, colourful, artistic, risk-taking, chaotic, random, intuitive, synthesizing, subjective, nonverbal; it processes feelings, imagination, symbols, fantasy; it looks at wholes.

The right brain is connected by the optic nerve to the left side of your body. It is the receiving, magnetic, negative, "in" half of your being. The female is represented by the circle.

According to Buys-Ballot's Law, a Law of Nature, when you stand in the Northern hemisphere with your back to the wind, the low-pressure system will always be on your left. This assists you in understanding that the left side is negative.

The female side is your creative intelligence. It is the doing, creating, building, and hands-on part of you.  The right brain includes activities such as drawing, dancing, visiting a museum, telling stories, laughing, role-playing and daydreaming.

## male

The left hemisphere of the brain is male.  It is factual, logical, linear, time-based, sequential, rational, analytical, objective, verbal; it deals in reason, facts, words, reality, safe practice; it looks at parts.

The left brain is connected by your optic nerve to the right side of your body.  It is the transmitting, dynamic, positive, "out" half of your being.  The male is represented by the cross.

The male side is your thinking intelligence.  It is the planning, rationalizing, designing, and theoretical part of you.  The left brain includes activities such as doing puzzles, reading, classifying things, writing, debating, doing math and spelling.

## hints for the wise

♦ It takes a woman and a man together to make a baby.  Just as it takes a Goddess and a God to make a world.  This is just common sense.

♦ Goddess has three aspects: Maiden, Mother, Crone.

♦ God has two aspects.  His spirit resides in the centre of the Universe, but you can know him through the Sun.  It is like his chariot.

♦ You can know Goddess through your relationship with your Mother.

♦ You can know God through your relationship with your Father.

♦ You cannot argue with the Laws of Nature.  They have been proven to be irrefutable.

♦ You can connect with the heartbeat of the Creatrix by meditating with your breath.

♦ It is wise to bring balance to the two sides of the body instead of favouring one side over the other.

♦ The Creatrix is inside of you.  You are a miniature Goddess or God.

♦♦♦

Two represents the balance of life.  It is the heartbeat between Goddess and God.  There is a pause between the switch of direction.  It is a moment between the push and pull which is the hidden third component.  You cannot have two without having a bridge, the third element in the exchange.  This is the beginning of the spiral, the spiral of life.

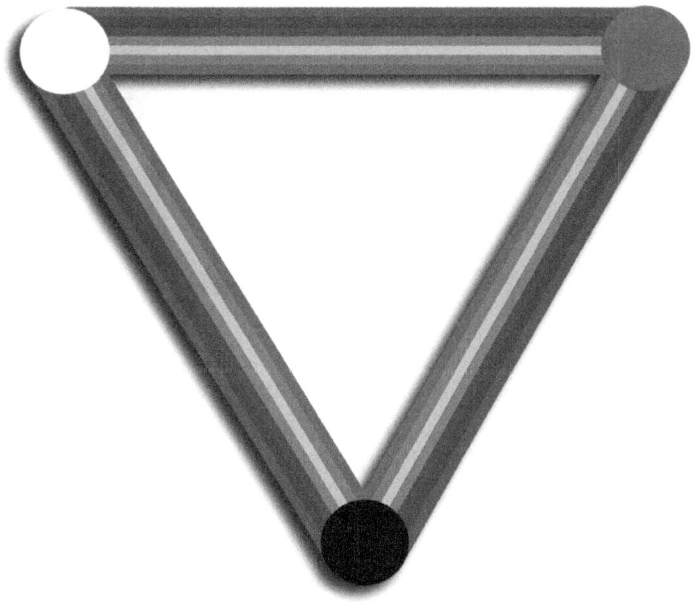



Three is the cycles of life in which you participate.  A triangle is the strongest geometric shape, a human ear has three canals, there are three primary colours, braids have three sections, atoms have three constituents, there are three essential principles of alchemy: salt, sulphur and mercury, and the Goddess has three aspects.

# trinity

The Goddess, the Earth underneath your feet, is alive.  She is a sentient being.  You know this because she grows just as you do.  She has a heartbeat.  She breathes.  She has blood.  The land is constantly growing, shifting and releasing.  She has many cycles just as you do.

There is only one Goddess.  She is infinitely vast.  Though you may learn about her through her many archetypes.  History paints her as many different goddesses.  She wears these different robes so you may know her.

The Goddess offers you her ancient archetype of the Trinity with her three symbolic aspects of Maiden, Mother and Crone.  She is a triple-deity.  Meaning she wears three different faces, though she is one being.

The concept of her different aspects speaks to you of the spiral of life with its birth, existence, death and its ability to recurse back onto itself, creating life anew.  These cycles are in everything and everything is in them.  Their patterns can be noticed throughout your reality.  Birth, death, rebirth.

This Trinity can be seen in many cultures from around the world.  It is found in the Celtic Goddess Morrigan, the Egyptian Goddess Qudshu-Astarte-Anat, the Irish Goddess Brighid, the Greek Goddess Hecate, and the Ancient Roman Goddess Diana Nemorensis.[2]

The Trinity is present in many myths and legends.  It is a theme that keeps reoccurring repeatedly throughout time.  It speaks of the flow of life and its constant growth, development, demise and its ability to resurrect.  Of cycles that lead to rhythms, rhythms that lead to patterns, patterns that lead to currents and currents that lead to aeons of time.

These are divine forces.  They are ancient energetic currents that

have flowed since before time.  You are affected by these currents of energy.
You have contact with them throughout your life.  At any given time, you
are learning the lessons of one or more of these aspects of the Goddess.

These cycles are the lifeblood of what it means to be human.  They
are the very core of life.  They are where you come from and where you
are going.  You only have to look to see these natural cycles all around you.
Everything will change, except these primordial and foundational cycles of
Nature that affect you on a primal level.

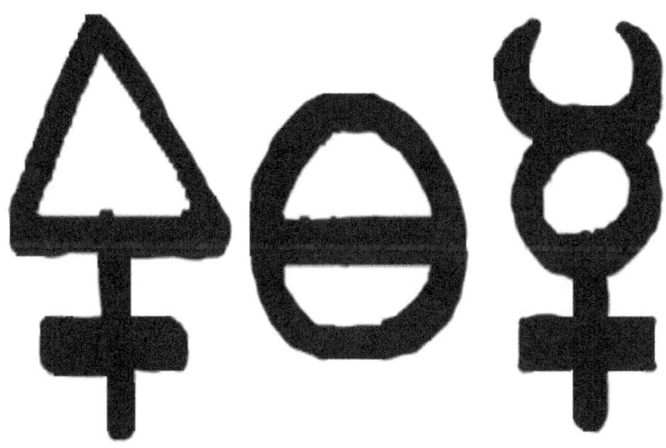

# 3 alchemical principles

Together sulfur, salt, and mercury represent the three alchemical
forms of energy.   They are the three prime essences of life.   It is not just
the physical interpretations of them that is important, but the philosophical
meanings of them as well.  They can bridge the gap between energy and
matter, philosophy and science, magick and reason.

Creation, preservation and destruction.  They represent the three
foundational properties of life in our Universe.  These fundamentals are the
process of life: birth, growth and death.  As well as the scientific principles
of combustible, matter and liquid; electron, neutron and proton; and oily,
mineral and alcohol.

The alchemists of old believed them to be holy.  They represent
the marriage of the White Queen with the Red King, in what was known
as the Alchemical Wedding.  The process in which birthed the mystical
Philosopher's Stone.

They are also the astrological concepts of cardinal, mutable and
fixed.  Hindu's call them the three gunas.  Ayurvedics call them vata, pitta,
and kapha. They are in everything and everything is in them. They represent
the turning of the Universal Wheel and the Great Hunt.

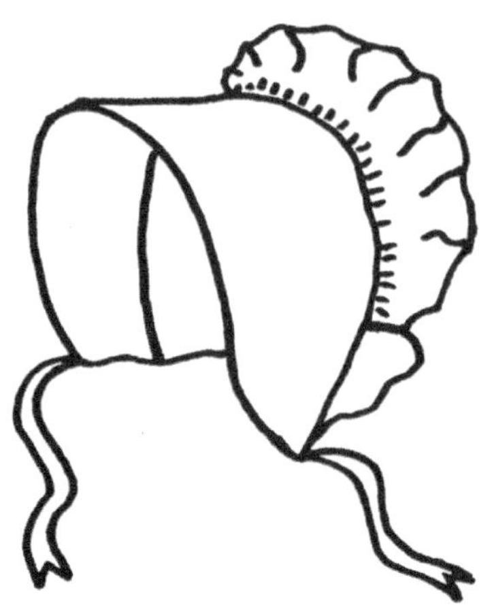

# maiden

The Maiden aspect is white, beginnings and potential.  It is positive, as in increasing.  This energy represents the childlike innocence and wonder you have when you begin your journey upon Earth.  It is a light, accepting and welcoming energy.

It is associated with an open heart, baby animals, gaiety, laughter, playing, dancing, being carefree, abundance of energy, spontaneous actions, virginity, snow, ideas, imagination, spring, a lamb, morning, and spring flowers.

She is known as the Maiden Goddess.  For men this would be referring to the Lad aspect of God.  She is represented by the new crescent moon.  This aspect of the Goddess represents birth.  All things born  come from a seed, an egg, or an initial beginning.  They are birthed, starting small, compact and vulnerable.  They have kinetic potential for growth.

# mother

The Mother aspect is red, creation and strength. It is neutral. This energy represents the stamina it takes to give birth, and taking care of things outside of yourself. It is a strong and nurturing energy.

It is associated with honesty, blood, protection, fertility, nurturing, community, sharing, sunlight, milk, work, altruism, creativity, manifestation, summer, a lion, evening, and tropical flowers.

She is known as Mother Goddess. For men this would be referring to the Father aspect of God. She is represented by the full moon. This aspect of the Goddess represents fulfillment. All things grow, develop and eventually mature into something different from how they began. They climax, fulfilling their role.

## crone

The Crone aspect is black, death and power. It is negative, as in diminished. This energy represents the wisdom gained with experience and age. It is a heavy, deep and direct energy.

She is dark because she is difficult to understand. The Crone is associated with wisdom, intuition, story-telling, the unknown, resting, boundaries, solitude, health, memory, endurance, far sight, loss, refuse, winter, scavenging, a serpent, nighttime, and dried flowers.

She is known as the Dark Goddess. For men this would be referring to the Sire aspect of God. She is represented by the dark moon, the three days before the moon is new, when the light of the Sun is not reflected upon its surface.

This aspect of the Goddess represents decline and death. All things eventually reach their demise, and begin to revert back to a simpler structure.

They degenerate and break down so that the parts may be reused in building something new again.  Death leading into rebirth.

The Crone does not represent just any kind of death.  She represents the pivotal point where deaths turns back into life.  Carbon into a diamond, rot into food, fossils into fuel, bones into nutrients, and compost into fertilizer.

mother father child

Another important way to look at the Trinity is with the union of mother, father and child.  Or Goddess, God and you.  This triad represents the ultimate form of balance.  When you have an expression of the magnetic balanced by the dynamic, the end result is manifestation.   You see this holy Trinity all around you.

Life on Earth is dependent upon this concept as procreation is how the human species survives.  It is also a process, and represents a fractal because of its ability to spiral throughout time.  You know this because you can trace your bloodline through your parents, grandparents and great-grandparents, all the way to the beginning of the human species.  It takes the Sun and the rain to make a rainbow.

the spiral of life

Every single living being, from a solar system, to a plant, to a cell, to animals and the parts of your body, grow according to the exact same spiral. This spiral of life can be found in everything natural that is around you and in you.

Cats, trees, your lungs, weather systems, and crystals are all spirals, and fractal. Fractal means that something is big and then descends into smaller parts that are repetitive miniature replicas. Some spirals just move more quickly than others do.

Nature has chosen this one thing to be the foundation of life. The spiral grows according to distinct measurements - a specific number sequence called the Fibonacci sequence: 0 1 1 2 3 5 8 13 21 34 55 89...[3] This is the ultimate Law of Nature. It is the fundamental template for existence.

You can physically see the spiral by watching water go down the drain, as it always spirals downwards. You can see the spiral in the tornado on the horizon. A giant storm forms a spiral on the weather channel. Seashells, solar systems, pea shoots and grapes that have little spiral arms, roses, snails and curly hair are all spirals. Maple seeds called helicopters fall spiraling to the ground.

Life is fractal. It is a eternal pattern that repeats itself. Everything living has small components that consistently grow to large and back to small again. Such as animals, which are found to be all sizes. They all start small at one point and grow larger, then get old and start to shrink and eventually turn into dust.

That dust becomes the minerals and nutrients that refuel another aspect of life in its perpetual cycle. Mountains turn into hills, just as branches turn into leaves. A baby animal riding its mother's back is also an example of something that is fractal.

The Earth and other planets spiral around the Sun. The weather and seasons constantly spiral upon themselves. Stars are born and they grow until they explosively expand and then they dramatically shrink becoming a new thing like a black hole.

Your emotions are fractal because they can go from small to expansive and back again quite quickly. You, yourself start as a tiny egg, and grow into a mature being that eventually diminishes back to minute particles. You can harness your spiral through your family. You can feel the cycle of life behind you through your parents and in front of you in your children.

The way to be aware of the spiral in your life is through your body, as it is a spiral.  Your fingertips are spirals.  The ratio between one finger and the next is based upon the spiral.  Your whole body is actually proportioned based upon this spiral.  Your heart grows like a spiral.

You can harness your spiral within by constantly changing your heart rate and actively engaging in life.  The more diversity of movement you have, the more vital health you can promote in yourself.  This means dancing, moving, exercise, balanced out with rest, relaxation and quiet time.  Your body knows how to ride the spiral.  It is just waiting for you to say, "Go!"

It means working and playing hard balanced out with serious relaxation to reach the heights of human potential.  A wise person knows how to relax the body with the mind.  You can paint away your body with your paintbrush of invisibility.  What a great way to erase stress.

The world around you is a constantly fluctuating spiral.  If you want to live a harmonious existence, you can welcome the spiral into your reality.  You can acknowledge it and follow its flow.

You can embrace the cycles you participate in within yourself that mimic the ebb and flow, the ferocious and quiet, the active and the passive energies of life.  You can find these expressions in your own life.  It is wise to accept them as your guide.

Three represents cycles.  It features the spiral of life, and how deeply its curvaceous nature reaches inside of you.

Four is learning the process of the circle that is a spiral.  There are four directions, four elements, four states of matter, four chambers of the heart, four blood groups, four fingers and toes, four seasons, four faces of the tetrahedron platonic solid, and humans have four distinct bodies.

## the four directions

From where you stand, there are four directions leading away from you: North, South, East and West. You are always standing at the crossroads. Each direction is its own way, its own path and its own energy. At any given time, you are facing one, coming from one, going to one, things are coming to you from one, or things will leave you heading into one.

The four directions can assist you in knowing where you are in the Universe. Not only in a physical sense, but also in a spiritual one. They teach you sensitivity because they show you how to become aware of your place.

Traditionally, there are certain energies assigned to each direction. For example, the West is where the Sun goes at night, so it is associated with endings and death. The Sun always rises in the East and so the East is considered the place of birthing and new beginnings. The wind from the North, the stories say, is always cold. The weather is hot in the South.

## the four elements

Humans since ancient times have acknowledged four elements as the four most fundamental powers in the Universe: Fire, Air, Earth and Water.  Your body and the Universe use these four simple powers as the foundation of everything.

The science of alchemy is based upon the elements and their relationships.  The elements have a symbiotic relationship with each other.  Fire needs Air as fuel, Water can extinguish Fire, and Earth contains Water.

When you eat breakfast you use the fridge (Earth, Fire and Air) to grab some eggs (Earth), you cook them in the pan (Earth and Fire), and drink some orange juice (Water).  Every single act of creation has its source in these primordial concepts of the four elements.  Every single thing in your world can be broken down into these simple energetic expressions.

Your body is an expression of the four elements.  You have the spark of life (Fire) that begets your heartbeat.  You breathe Air.  You are made of the elements that comprise Earth.  You have blood, which is a liquid like Water.  Being aware of the four elements is a beginning to establishing a relationship with the Earth.

The four elements have energetic properties as well as physical ones.  You can bring balance to your energy by bringing balance to the four elements in your life.  If something feels out of whack, you can relate it to an element.  Then intentionally focus on the other elements by bringing them into your life in order to righten any imbalances.  Elemental energies are potent and can be called upon for assistance at any time.

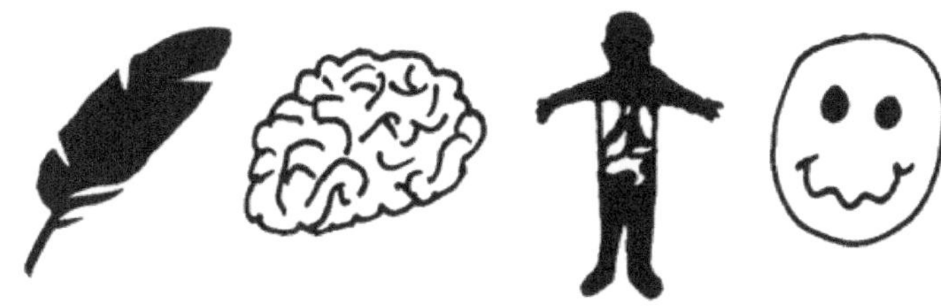

## your four bodies

You have four separate bodies that comprise you as a whole: your Spirit Body, Mental Body, Physical Body and Emotional Body.  Each of your bodies is a part of you.  These four parts of you make up the whole.

Each body is an aspect of you.  You can have different levels of vital health in each.  You may have a strong Physical Body with a feeble Spirit Body.  You may have an overpowering Mental Body with a delicate Physical Body.  You may have a mighty Emotional Body with weakness in your Mental and Physical Bodies.

The way to become a fully activated, immortal human being, capable of easily learning and assimilating all of your lessons in this lifetime, is by being equally balanced in all four of your bodies. The way to achieve this level of balance is through awareness of your four bodies. Awareness leads to authority.

## your four tools

It is wise to realize you have four tools: your Wand, Knife, Shield and Cauldron. They are both physical and energetic. They can assist in your personal evolution. You have an ability to create, possess and use these tools. They are there to give you choice and power to maintain command over your vital health and well-being.

Your tools are the physical manifestation of your ability to claim dominion of the four directions, elements, and four bodies. The tool is the representation of the authority you have with your sacred work. It is granted to you by your initiation of that quadrant of the circle.

Your four tools are special. They should only be used for sacred work. It is wise for you to make them yourself. Use tools that speak to you and are powerful for you.

## bring it together

You can merge these concepts of the four directions, the four elements, your four bodies and your four tools. There is a direct relationship between them. They come together harmoniously. You combine these concepts because they form the foundation for what is known as a circle.

It is wise to bring these four concepts together because it enables you to understand them, and therefore yourself more easily. Combining the direction, element, body and tool makes each quadrant of the circle an energetic construct that can be harnessed in order to bring awareness and ultimately power to you.

You can learn the subtle nature of each one of your bodies through its correspondences. You can understand your own energetic nature much more easily by assigning these primordial powers to them. These correspondences are symbolic in nature and incredibly deep in meaning. It is wise to ponder them.

The correspondences of the circle are:

♦ North, Fire, Spirit Body, and Wand.

♦ South, Air, Mental Body, and Knife.

♦ East, Earth, Physical Body, and Shield.

♦ West, Water, Emotional Body, and Cauldron.[4]

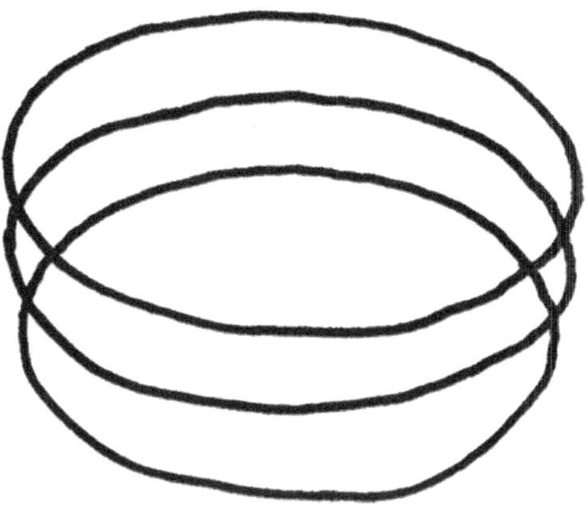

## what is circle

A key concept to being a real human is to know where you are in the Universe. A great way to do this is by the ancient technique of casting a circle. It is a way to acknowledge the seven directions as they relate to you: North, South, East, West, Goddess below, God above, and within your heart as the seventh direction. This brings awareness to the place in the Universe in which you reside, enabling you to claim dominion of it.

You create a circle by acknowledging the boundary around you. You mark this space with a leather cord. This cord energetically represents the periphery of your circle. It is a symbolic initiation of reining in your sparkle by harnessing your four directions, elements and bodies.

You claim dominion over them with one of your four tools. In this way, you develop a working relationship with the energetic world. You acknowledge that it exists by claiming dominion of your place in it.

Within your circle, you place your four altars. There you lay your directional tools and mementos that represent that specific direction. These altars are physical representations of the direction that hold the power of that quadrant. They take on life as they become empowered with the energy of your spiritual circle work.

By casting a circle, sacred space is created around you so you may easily connect with the Creatrix. This is a vortex of energy. It is a place outside of normal time and space. It is a whirlwind of essences that wrap around you. It makes everything you say or do take on power, as it connects you energetically to Goddess below and God above, to the four directions, the four elements, your four bodies and your heart.

Your circle is a place of communication and empowerment because it concentrates your energy in one place, instead of having your essences spread haphazardly. It is a sparkle storehouse because your circle grows stronger every time you call it. By having clear and definable boundaries, your energy is safeguarded from being drained and depleted. It creates atmosphere that is completely outside of normal reality. It is perceivable and evident.

Casting a circle energetically defines your boundaries and empowers them with Creatrix energy. This active boundary magnifies and focuses your attention. It heightens your sensitivity to your life experiences. It brings you awareness to your parts. By containing your energy for a time, you enhance it. The circle acts like a battery of energy fresh from the Creatrix, naturally reviving you.

There are no rules to life. Live it as you see fit. Goddess and God are neutral and accept you no matter what you do. There is no punishment system or judgment system governing humans.  Whatever you put into your circle will come back to you.  That is the underlying foundation of the circle and for life as well.  So it is wise to choose carefully.

Casting a circle is like a rollercoaster ride at the amusement park.  It can be scary at first because it is unknown.  Once you get over that fear and realize that circle is an exciting ride, it becomes fun.  It is there to teach you how to access your own inherent powers.  It enables you to enhance your connection to your own vital energy.  What a thrill.

## Hints for the wise

- When something comes to you, it is being affected by the energy of the direction from whence it comes.

- Life is elemental.  You can relate everything in your life to the elements communicating to you.

- It is wise to become aware of this and figure out ways to use all of your bodies equally.

- You can create altars all over your house by bringing Nature indoors. Nooks and crannies work well.

- You use tools all the time - why not spiritual tools? It is wise to find ways to explore your relationship with the elements.  Discover where you already interact with them.  You can figure out ways to be involved with them in new and exciting ways.  Bringing awareness by acknowledging them can do wonders for your soul.

    ♦    At any time you are acting from one or more of your bodies.

Four represents the four directions, four elements, four bodies and four tools of the human that come together to form the circle.

Five is being a spirit person and the way to do so.  The elliptical path of Venus forms a perfect pentagram, an apple cut sideways has a star inside, sand dollars have a star on them, flowers have five petals, starfish have five arms, and humans have five basic senses and five digits on each extremity.

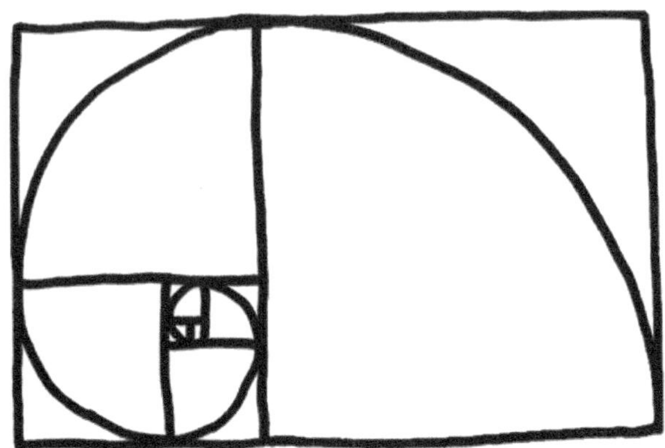

## golden ratio

The pentagram represents the golden ratio of 1.618.[5]  This ratio represents the spiral of life, or the Fibonacci sequence.  The official blueprint used by every single living being as it grows.  It is the ratio between the proportions of the spiral, represented by the different lengths of your fingers.

This ratio has been used for centuries in design, mathematics, architecture, music, and in painting.  Artists find art based on this ratio to be the most agreeable and satisfying for the human eye.

## the platonic solids

The platonic solids, five in number, are the building blocks of the Universe.  These shapes are the foundation of the physical Universe. Development of everything in creation follows them.  They are part of the science known as sacred geometry.  They are the building blocks of growth for creation.  Every single thing in the Universe can be represented by sacred geometry and its harmonic proportions.

Sacred geometry represents harmony.  It demonstrates a method of the divine interacting with humans on the energetic plane, as well as a physical one. It offers humans subtle guidelines that could be used as the foundation for life. Including the construction of  modern systems, structures and designs.  Mimicking Nature and her Laws brings joy and bliss to humans.  It is wise to bring awareness to the sacred geometry in your life as much as possible.

## spirit person

The star or pentagram represents the five components of a spirit person.  The soul is the fifth element that has claimed dominion of the four elements and creates the star.  A person of spirit is someone who has worked diligently at maintaining a relationship with their soul.   The star is the emblem of an ensouled spirit person.

For example, when you regularly do circle, meditate or practice being in balance, you meet with your soul regularly.  Therefore, it trusts you and lives inside of you.  This makes you a spirit person.  They can be recognized by the sparkle in their eye.

A spirit person develops their four bodies by focusing on their personal balance and harmony.  They are fueled by a solid connection with the Creatrix.  They know themselves.  They are able to communicate directly exactly what they mean.  The false things in their lives have simply fallen away.  They are usually patient and consistent.

Humans are amazing.  When you give yourself the freedom to aspire to your true potential, you blossom like a rose.  You have so many talents and skills you probably do not even know about.  You have a passionate spirit that wants to shine.  You possess an amazing memory and a powerful mind that wants to learn and expand itself.  You have a body that can accomplish incredible things, if given the chance.  You have emotions that can sail you to the moon.  Humans can have an unending amount of stamina, wildness and creativity.

Every single human has the potential to be a spirit person, whatever that means for them.  Some just do not know it.  Every human is, to some degree, telepathic, empathic and able to read, see and understand energy.  This is your natural way of being.  Spiritual evolution comes from actually doing the things you think about doing.  You can clean, organize, get into the body, move, focus, and work diligently to improve the quality of your life.

When you are one with your soul, you realize it is eternal and immortal.  It is inside of you.  It enables you to let go of fear, reason and your limits.  The Will is the vehicle that drives the soul.  The spirit person walks with their Spirit Body first.  They interpret reality through their spiritual vision.  They wear their heart on their sleeve.

Spirit people tend to be quirky; they stand out from the crowd.  Their energy is perceivable and tangible.  They love to express themselves fully.  Or, they are hermits whom you never see.  Spirit people can dress differently than normal people.  They can wear bright colours or all black.  They can have cool gadgets, beautiful jewelry and nice accessories.  They are usually smart and wise.  They constantly seem to say something you would not expect.

Spirit people usually have not had an easy life.  The spiritual path can be a rough one.  Developing the character of a spirit person has many difficult lessons.  The spiritual life has severe periods of loneliness balanced with consuming periods of intensity.  This is the way it must be.  Through the loneliness, you carve out your strong character.  People of spirit can tend to be loners.  Or, focused on tribe and clan.

A spirit person has usually surpassed the angry headspace of blaming others for a difficult life.  They have realized their skills, talents and powers were worth the pain.  Spiritual people usually have had a difficult upbringing for many different reasons.  It is difficult in order to develop strength.  Before your birth, your soul chooses your parents and the lessons involved, in order for it to grow.  The lessons were tough because you needed them to be, so you could become more of who you are.

Spirit people have a certain glow.  They are passionate.  They have a certain glint and knowingness about them.  They are magnetic.  They usually make you feel vibrant when you are around them.  Or, they can be quite scary.  You may want to avoid them.  Sometimes they say things that can haunt you for the rest of your life.  Spirit people tend to say the thing that etiquette says should not be said.

Spirit people usually do not like to conform. They are pioneers. Spirit exists outside of the labels. Spirit people tend to be different. They naturally do not fit in with everything around them. They do not need rules because they are their own leader. They tend to shy away from collaboration. They can be stubborn and insist on doing things their own way. They are interested in quality and originality.

Soul can change your life and the people in your life. When you are a spirit person, you usually reach a point where you want like-minded people in your life. Those who are not keeping up with your changes seem less appealing as compared with people who themselves are on a spiritual path.

Spirit people can have intense energy. When they get around Muggles, they can trigger the Muggle into starting to feel intense without knowing why. The spirit person is showing them how to use their own innate spiritual skills, just by being around them.

The spiritual path places the soul above everything. When your intention is to walk with your soul, your path becomes harmonious. The soul inherently is a grand fountain of joy. It loves you. It wants to keep you safe and in bliss. Therefore, it works ceaselessly to engage you. Your soul is your armour. It will initiate you into the guild of the Spiritual Warrior.

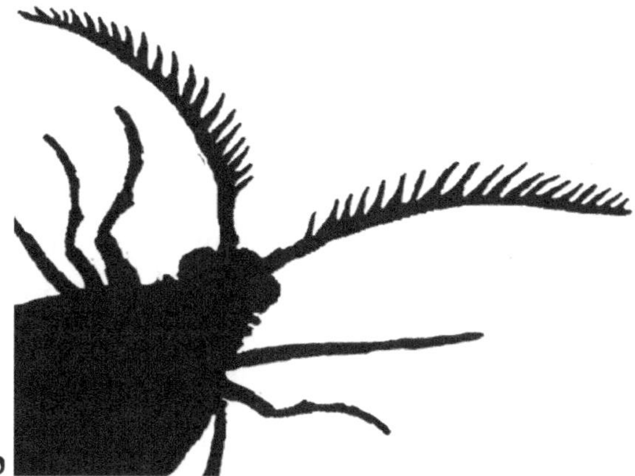

## awareness and sensitivity

Awareness is a key to becoming a person of spirit. Awareness is being conscious of your surroundings. As a spirit person, you evolve into having finely tuned senses. Sensitivity is a personal experience. It is different for everyone. It is the skills and talents you personally acquire. It is also the subtle and nearly unexplainable awareness you discover about yourself. This knowing is a source of your sparkle power.

Awareness is allowing yourself to feel your surroundings. You take in the signals offered to you by the things in your periphery. The living creatures around you are constantly sharing with each other their interpretations of the environment. For example, following the Laws of Nature, animals work together to be safe. They will warn each other of any approaching predator. Birds, no matter the species, team up when a cat approaches.

It is no different for humans. You have the ability to be aware of your environment and the subtle messages it is sending your way. Awareness is the art of deduction. It is assimilating the hidden signals things are giving off. You can use these signals to educate yourself about the true nature of your environment.

Being sensitive is having a keen ability to communicate with the things around you. It is learning to be sensitive to the plants, animals, energy, emotions, people, situations, minerals, auras, environments, and objects in your life. Everything around you is in a constant state of communication. You just have to tune into it.

This subtle form of communication leads to the ability to be in command of your reality. It gives you the upper hand in knowing your environment. By being intimately aware of your surroundings, you become empowered. You have clarity instead of uncertainty and can easily recognize what you need, when you need it, instead of struggling to figure things out.

Becoming sensitive to others is learning love free of conditions. It is about understanding how to become more empathic towards others. An empath feels the emotions of others. This tends to lead towards greater compassion for humanity. When you feel others, it enables you to comprehend them more easily. This leads to a greater ability to accept them. It leaves you less inclined to fuck with them.

Being sensitive comes with a price because you have to learn to shield yourself. The pain of others and the pain of the Earth can overwhelm you. Spiritual pain is different from physical pain because it just is, and you feel it way more deeply. The world is hurting, and sensitive people pick up on it. There are many sources of spiritual pain, including nuclear war, trees being cut down, harm to animals, too many crazy people, and pollution, for example.

So many people are numb to the pain. As a sensitive, you fully feel it. You allow yourself to. The intensity of being sensitive and feeling the pain of the world is a form of initiation. As a spirit person, you become used to it and learn to carry on. Vulnerability takes way more strength than being an asshole ever could.

As a person of spirit, experiencing spiritual pain and then releasing it, you are assisting the Universe in clearing the energy of trauma. It goes through you and into the Earth, where it is processed and turned into energy that is more viable. In other words, you make it disappear. This task is incredibly powerful. It is humbling because of the long-lasting effects it has on the world and your karma.

Unfortunately, there are many humans who are asleep. They are unaware of their own consciousness. They are unaware of their spirit selves. They are unaware of the living energetic world around them. They are unaware that the Earth and her creatures are sentient and alive. This unawareness can lead to many adverse choices that can greatly impact the world around them.

When you are spiritually asleep, you are energetically ignorant. This ignorance can lead to creating karma because you cause harm to alive and sentient creatures. Ignorance does not protect you from Universal justice. You will pay for your actions of harm, whether you are aware of creating them or not. Being asleep is like living in a tiny box. You think you are alone and it does not matter what you do.

Sensitivity is awareness of your environment. It is your feelers, your antenna, your tentacles, your alertness, your perception, your keenness, your understanding, your cunning, your intuition, your discrimination, your premonitions, your hunches, your instinct, your inspiration, your sixth sense and your whiskers. In other words, sensitivity is the root of all of your communications with the outside world.

sentiment

An important power you have as a spirit person is your sentiment. Sentiment is different from intention because your intentions shift and slide. Your sentiment is in your heart, where it stays.

Acceptance is your armour as a person of spirit. Acceptance of the world the way it is. Not the way you would like it to be. You can allow innocence to be your guide. It is not naïve innocence. It is not based upon ignorance or apathy. It is the innocence of a fully aware being who consciously chooses vulnerability as their shield.

Vulnerability means you are open to receiving, as well as to sharing. In this way, soul flourishes. This is what it means to have an open heart and an open mind. Being open shows others how to do the same.

Sentiment is your true direction and purpose in life. You can do all of these things recommended in this book, but if you do not have the proper sentiment, nothing will work for you. Gratitude is an important attitude to have while pursuing the spiritual path. By focusing on what you are grateful for, you will manifest more of it.

Feeling as if the world owes you something, having a chip on your shoulder or having expectations of the Creatrix are all sentiments that can eventually fall short. There is a balance in your sentiment that cannot be taught or learned. Karma will ultimately teach you this.

As a spirit person, you can brim with vitality and awareness. You can have a strong connection to the Earth and the creatures in which you share your space. You have gratitude for being alive and acknowledge yourself as a small cell in the body of the ecosystem around you, celebrating the interconnectedness of life on Earth.

To be a spirit person is to strive ever towards aspirations of a world that is balanced, holistic and in harmony. You can integrate with Nature and work towards becoming one with the Creatrix. There is much more to being human than is expressed today.

There is much more to be learned, accomplished and self-realized. Collectively, humans can break through the illusion of impotence and realize our creatorship. The bridge to your sentiment would be your memories of when you were five or six years old. What did you do for fun? What were your favourite games? What made you content?

Sentiment is about wonder. If you have lost your wonder for life then it makes living more difficult than it has to be. It is wise to make wonder a priority. In this way you are safeguarding your relationship with your soul. Wonder is your soul's natural way.

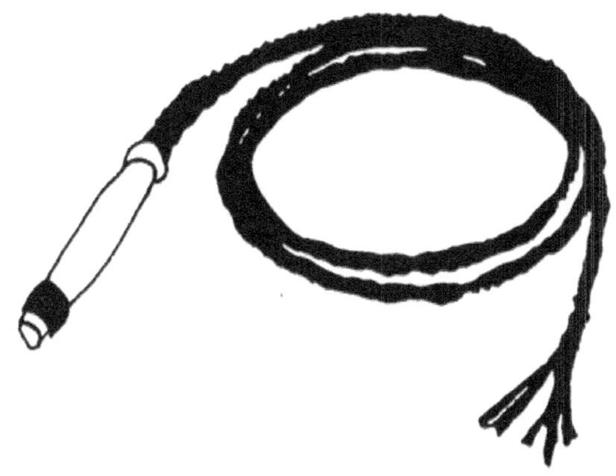

# discipline

What does it take to become a spirit person?  Discipline.  Being a spirit person requires plenty of discipline because strengthening yourself requires time, effort and patience.  It is like the tempering of a blade.  They are lessons that are painful exercises in endurance, strength and stamina, in order to develop character.  There are also marvelous heights to be reached when you accomplish your goals that balance out all the hard work.

Discipline is the iron spine that traverses your spiritual career.  Spirit people follow a code that they define for themselves.  This is what separates spirit people from non-spirit people.  Discipline in spiritual endeavours is how you get, grow and maintain your soul.  It takes constant stamina and endurance to be a person of spirit.  Spirit people tend to take their work very seriously.

Discipline is painful.  It is the pain of growth as you transcend your shell, expand your personal consciousness and learn about yourself.  Discipline includes spiritual, mental, physical and emotional control.  Not letting any one area overpower the rest.  You learn how to monitor yourself.  When you do not like what you are experiencing you change it, by changing your behaviour.  Discipline means you work at consistently performing at the highest level.  It means you continually focus upon improving your skills.

No one is coming to save you.  You must save yourself.  There is no higher power interfering with your reality.  You have to work diligently at bringing that higher power to you.  You have to work at empowering yourself and ensouling yourself because it will not just spontaneously happen.  Discipline may be difficult, but in the end it is strangely addictive as you get accustomed to the feelings of awesomeness it brings about.

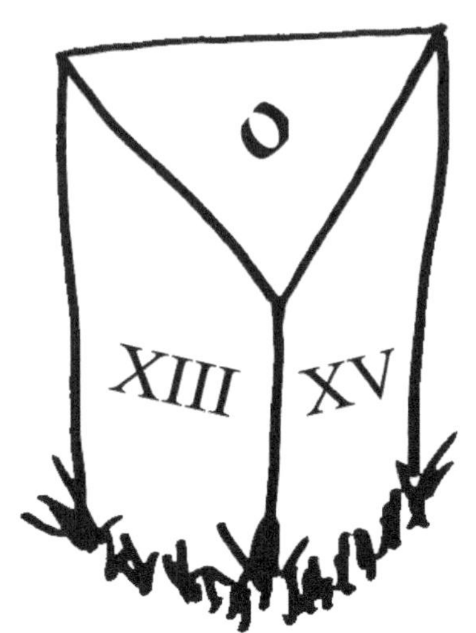


## boundaries

Being a person of spirit means you have defined the boundary of your circle. You know that having and maintaining personal boundaries is a way to success. Boundaries are a way to empower whatever it is you do. Concentrating the energy will always bring results.

Boundaries are personal rules. To have and maintain personal boundaries, you must first be aware of what they are. You must know your limits, your desires, your abilities and know how you feel. Honouring your boundaries means you know your capabilities. You know how far you are willing to go.

Having personal boundaries is saying no. It is keeping yourself together. It is honouring yourself first. It is not doing something because you do not want to. It is doing things because you want to. It means you do not let your shit fall out.

Having boundaries means you do not always have to share what you know. You do not leak energy by talking too much to friends about the state of your life. You can lose a lot of energy by over-talking, as it is a weakened boundary.

A feeble boundary, like a thin aura, leaks energy through it. It is wise to limit the number of times you tell stories and the number of people with which you share.

Maintaining and strengthening personal boundaries teaches you how to respect other people's boundaries. Respect must go both ways. It is not pursuing or attacking other people's boundaries because of your own weaknesses. Instead, you have your own strength in your core to maintain your own boundary. You have no need to interfere with others. Healthy boundaries can be taught by leading by example.

## trust

A spirit person has learned to trust. They know when they reach out their hand, the tool they want will be there to grab. Trust comes from practice and a willingness to surrender. It is the enthusiasm to be led harry-merry down the garden path in a state of clarity and confidence.

When you are in service to the self or the ego, you try to control things with your mind. If this is the way for you, then go for it. You will only get as far as your mind is able to define.

Instead, when you are in service to the Creatrix you trust in the way made clear for you. Suddenly your options become unlimited because the Creatrix is able to imagine an infinite world. This limitless way can become your way, if you allow the Creatrix to guide you.

Things you would never have dreamed of can start to manifest in your life. They can truly bring you bliss. Are you running out of resources during a project? Just know there can be enough and there will be. Say you cannot find what you are looking for in a drawer. Maybe looking one more time with an open and trusting heart, you may be surprised when you find it.

Trust is surrendering to something greater than you are. At first, it may seem difficult. It just takes practice. A certain peace comes when you surrender to spirit. It means you wear your vulnerability as armour. It opens you up, like a geode sharing its sparkling crystals. At the same time, it brings you inner strength you cannot find anywhere else.

Trust begins with self-trust. You can learn to listen and act upon your own intuition and instincts. It becomes easy to realize that you are trustworthy and actually know what the best thing for yourself is, more than anyone else.

## courage

To be a spirit person is to be brave. They have learned to not be afraid and turn that key. Spirit people have learned to deal with their personal freak-outs. They still accept themselves and walk proudly, head held high. They walk to the beat of a different drum.

You can learn that these periods of intense emotions are the current of the Creatrix trying to express itself through you. There is so much quiet, controlled and repressed energy that is held back in society. Spirit people are channels for releasing that pent-up, frustrated spiritual energy of reality. They are energetic pressure valves.

Power is sometimes not an easy thing. It separates you from others because it makes you a leader. It makes everything you do have a heightened impact. When a spirit person speaks, people listen. It is wise to be damned sure you know what to say. The responsibility of being a natural leader to others takes courage because you instantly feel the repercussions of your decisions in a way that followers just do not understand.

Strength of a true spirit person does not come from sacrifice or suffering. It comes from conquering their fear. Spirit people have walked through their fear. They have offered themselves up to the cosmic powers, whatever that means for them. When you ride the flow from the Earth to the ethers, it changes you. It makes you different from who you were and from others.

For example, it is a great day on the spiritual path, when you realize the voices in your head are other beings trying to communicate with you. It takes a subtle ear to differentiate the voices of the ancestors, crystals, elements, energy and the spirit world, as they shift in the layers of your mind. It is a mark of bravery because being a person of spirit means you do not simply conform to what society deems as correct. Instead, you make your own way.

Most people fall into a settling pattern. They are probably not blissful, but are too frightened, ignorant or lazy to propel themselves forward. So, they settle for what they have. This kind of compromise in life makes for a shallow vibrational existence. You are not fueled by bliss. Instead, you have a sort of bitterness that propels you forward. This kind of propellant causes calcification and aging of the body. It is wise to learn to face your fears and walk through them. The courage will come after.

Risk-taking is a way to propel you quickly through life. It is fun and exhilarating. It challenges you to stretch your ability to believe and is highly rewarding. Courage is self-confidence. How you feel about yourself directly reflects upon how you look to others. Your Emotional Body literally carves out and dictates your Physical Body. You create your Physical Body by how you feel about yourself. Walking through your fear creates beauty on a physical and soul level.

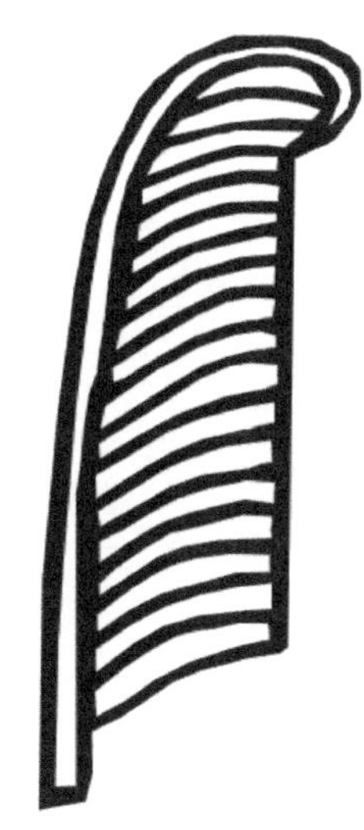

## truth

Truth is a real thing that shines. It is not bendable, malleable or inconsistent. It is hard and bright. It is measurable. Truth is discernible from non-truth.  There are many different ways of interpreting truth.  But, these considerations do not alter the validity or the original condition of the truth in any way.

As a spirit person, the truth is your guide.  The more you honour and be a proponent of the truth, the brighter you shine and ultimately the more powerful you become.  Truth is the most powerful thing in the world.

The truth can be painful.  In fact, it can suck.  There is no denying it.  There is no way to escape from it.  There is no way to avoid it.  The truth will always come to light.  It will always surface, like oil separating from water.  The truth can release you from pain.  And, like the cheesy saying goes, it shall set you free.

Acknowledging the truth makes you stronger.  You need to discover the truth about any situation and then go with it because it flows with Nature.  When you have the truth on your side, nothing can rock you.  The truth is flexible because it is always expanding and spreading.  You can always learn more truth because it is always growing.

Most of the world's problems are caused by lying, holding back and censoring the truth.  So much trouble and bullshit is caused by people not fessing up to what is really going on.  Humans could save so much time and energy by realizing it will come to light anyway and so they might as well share the truth.

## studying

The plain truth is, spirit people study their asses off.  They read, read, and read some more.  They take courses, they discuss as much as they can with their elders, and they learn from every source available.  They take notes.  They watch documentaries.  They go to workshops, conferences, and lectures.

Spirit people think.  They contemplate.  They takes notes.  They memorize, and memorize and memorize. They join study groups and have long discussions into the night. They have friends who also study, in order to bounce shit off each other. It is hard work to be a person of spirit.

Being a spirit person means making a commitment to yourself to be the most that you can be. It takes constant learning, studying, practicing and devotion to pass your spiritual tests. That means not getting off the train by deciding you have reached the end of your learning.

Light is information. When you learn, you are revealing more space in your mind with this light. It makes your personal mindscape bigger. This is what it means to have an open mind. This directly affects others.

Studying and reading the works of others assists you in developing your own beliefs. Agreeing or disagreeing with the opinions of others makes your mindscape ebb and flow with growth as you take in the light of information. It can fire thoughts in your brain that lead you down roads you would not have discovered on your own.

## spirit name

When you declare yourself a spirit person, you usually want to call yourself something to celebrate this fact. When this happens, you reach a point where soul becomes more important than being normal. Your priorities shift. You want to share your real self with the world.

You can declare yourself publicly by telling your friends and loved ones that you are a spirit person. From now on you will not compromise your beliefs for anything. This is an empowering day. Like a kind of birthday.

A form of initiation when you become a spirit person is to find yourself a name. This is your spirit name. It usually just comes to you. You can go so far as to change your name officially with your family and friends, and even put it on your driver's license.

Your spirit name represents a shift from a normal you to a spiritual you. It is also a shift in your personal vibration. You carry and communicate the vibration of your spirit name. When you want to show the world you identify with your bliss lightning in an intense way, you are ready for your spirit name. It is an empowering time because it makes you feel complete. You have found something in life that is truly important to you. Your spirit name brings power and self-confidence.

Witch, spiritual master, magickian, priestess, wizard, enchantress, warlock, conjurer, sensitive, enchanter, ipsissimus, wise person, pellar, medicine person, necromancer, solitary practitioner, guru, teacher, mystic, occultist, sorceress, mage, exorcist, shaman, medium, witch doctor, Nature worshipper, conjurer, guru, priest, having the shining, archimage, touched, fortune teller, seer, alchemist, clairvoyant, magus, oracle, sibyl, cunning woman or man, soothsayer, sorcerer, augerer, wise person, or diviner are some common names that a person of spirit can call themselves.

# alphas

An alpha is the one with the most sparkle. Nature works with the alpha female and the alpha male running the pack. It is no different with humans. When you become a spirit person, you are strengthening yourself. Odds are, you will be leading the herd because you are the most powerful person around.

Alphas are the only law. What the alpha says and does is the only way. This is a Law of Nature and the way of Sacred Law. This eradicates the need for manners and etiquette. It makes things simple. The power of the alpha is neutral. It is a power that is granted by Nature, as Nature is neutral.

The alpha is the most powerful person in the room. Not elder, the person with the most money or the smartest. Power dictates the alpha rules, not the elder. Being truly powerful is not about threatening or harming others. It is not about intentionally overpowering others. It is not a matter of destroying things or taking away power from others.

Real power is about your ability to be a leader. It is about being a solid role model and leading by example. Power is about doing the best thing. It is about maintaining a code of honour. Whatever that means for you. The alpha of the wolf pack will always send in the omega first, to test out the waters. The weakest link is meant to be disposable. That is why they are weak. They do not have the strength within their genetics to survive. This is how the Goddess corrects herself via sickness and disease. It keeps the genetic lines strong, to sacrifice the weakest link.

In society, the needs of the omega are put before the alpha, which is directly running counter to the Laws of Nature. Allowing the omega to rule the pack diminishes the ability to be strong and weakens the genetics of the human species.

Five represents the ensouled spirit person as a star reining over their four bodies, elements, tools and directions. A bee takes pollen from a flower (pentagram - five) and stores his honey in the hive (hexagon - six).[6]

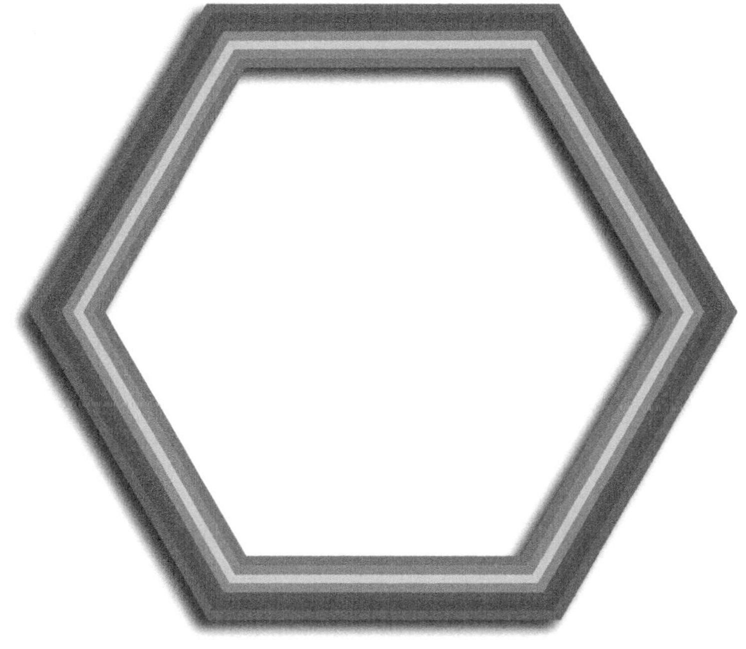



Six is your personal alchemy. It is everything in, on and around you. Snowflakes have a six-fold symmetry, there are six faces of the hexahedron platonic solid, insects have six legs and the honeycombs of bees are six-sided.

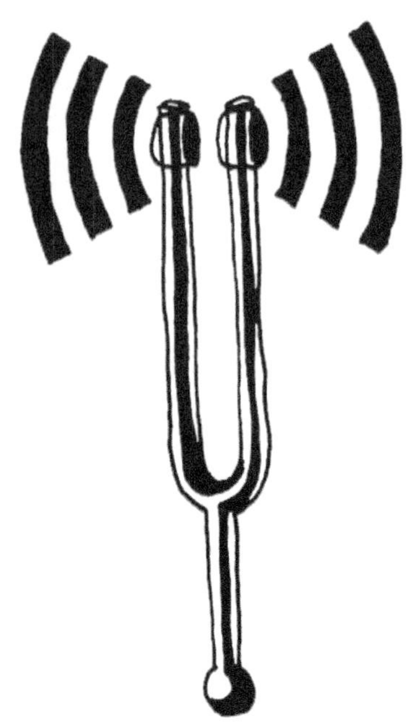

# vibration

Everything in the Universe has a vibration. Vibration is the energetic signature that something emanates as the energy is passed back and forth between the electro and magnetic subatomic components within. This back and forth is the heartbeat of the Creatrix. Everything is an expression of it.

Some things are closer to the Creatrix. They have lots of space for the heartbeat of the Creatrix to exist within. Therefore, they vibrate more intensely and have more sparkle. Their proximity to the Creatrix makes it easy for them to mimic this vibration.

Other things are farther away from the Creatrix. They have less room to hold the heartbeat of the Creatrix. Therefore, they have a weaker vibration and less sparkle. Because of the distance from the Creatrix, they find it more difficult to mimic that vibration.

Vibration represents the energetic constitution of something. Everything in the Universe is made up of energy. Some vibrating intensely, some vibrating less intensely. Everything has a vibration. All people, places, inanimate objects, weather, animals, experiences, memories, plants and planets, literally everything vibrates.

To be close to the Creatrix something has to have been birthed by the Creatrix. They are organic by nature and have not been fiddled with. Things farthest from the Creatrix have been fabricated. They are inorganic in nature and have been processed, usually by machines.

Every one of your alchemical choices changes you on a molecular level. You can either choose to evolve yourself and increase your sparkle; or, de-evolve yourself, and decrease your sparkle. You are born like a star. You only have so much energy and so much sparkle for this lifetime. Every single little thing you do adds up. What you consume will either assist you or deplete your natural energy.

Everything in your life carries vibrations that create and affect your well-being and health. Including the food you eat, the clothes you wear, the products you use, and the environment you live in. It is wise to have your alchemical intake be of the purest vibration that you can make it. By doing this, you maintain a powerful vibrational existence. This leads to a robust quality of life. It equals vibrant health and a long, blissful life.

Your personal vibration consists of layers and layers of choices, each piled upon the last. These choices can ultimately create vital health or energetic blockages. You can always change your mind and go in another direction. You just have to make different choices.

Vibration can be intense, weak or somewhere in between. For example, on the one hand you have a loaf of wonder bread, made by a machine in a factory. On the other hand, you have a loaf of bread made by your grandma from grains she grew herself.

What vibrational signature do these two loaves of bread have? Food made with love by someone you know, has an intense vibration and way more sparkle power. Processed foods with questionable ingredients made by machines tend to have a much weaker vibration. Vibewise, they have diluted sparkle power.

There is a monumental difference between something the Creatrix made, and something that is fabricated. A rainbow in the sky is an entirely different creature than a gasoline rainbow. It is a case of real rainbows versus unkind rainbows. Humans have not and never will figure out how to manufacture sparkle.

Because everything has a vibration, there is the matter of vibrations meeting and coming together. Some things come together well and resonate with each other, while some things do not come together well. They do not resonate with each other.

As you are of a certain vibration, you will notice that certain things resonate with you. When certain foods, people, types of fabric, and places make you feel blissful, or resonate with you, go with it. When things do not resonate with you, or they do not make you feel blissful, then avoid them.

Vibration communicates this way to you. When your vibration matches another vibration, you will feel energetic, excited or stimulated. It is wise to follow your feelings when this happens. This is your bliss communicating with you.

Because you are magnetic like a magnet, you have the ability to attract things, but also to repel them. This natural ebb and flow of energy around you can be used as a guide in your life. When things feel favourable and you are attracted to them, go with it. When things do not feel right and are repellent, avoid them.

Resonance is also a great way to discern information or ideas that you come across. If you read it, feel it and it resonates with you, then you know it is true for you. Feel free to trust yourself and believe it. If something that you read immediately sounds alarms in your mind, then just let it go. In this way, you are able to discriminate wisely between things in your life.

All matter is energy. All energy has a vibration. Vibrations are true. Your senses can readily pick up on them and interpret them. When you develop your sensitivities as a spirit person, you get adept at sensing the vibrations of things, people and places. With practice, you can pick up on the true vibrational value of things. They may not necessarily be the same as presented on the surface. Your senses can discern what lay underneath and reveal the true nature of reality.

The first step in becoming conscious of your vibration is to bring awareness to what you buy. It is wise to monitor constantly your intake and what you purchase. Your buying dollars are the most political decision you make every day. They are intimately connected to your personal vibration and alchemy.

The grocery store contains roughly 65% shit. Processed food, sugar, flour, stimulants, white powder, and artificial foods are not real food. Unfortunately, these things can give you a false sense of feeling good when you consume them. When in reality they are detrimental to your vital health.

You have to learn to weave your way craftily through reality, in order to maintain your vital essence. It is wise to realize corporations are using wood fiber, plastic, chemicals, artificial crap, and carcinogenic ingredients in their food to stretch it out and increase their profits. With practice, it becomes easy to ignore all the colourful packaging and focus on real nutrition.

# secret alchemical formula

The secret to life is alchemical. You have your own personal secret alchemical formula. It is the exact combination of things you can consume that is best for you and that will empower your vital health. It is wise to find it. It includes many alchemical keys that unlock the real you. Everyone has a different set of keys and a different way of consumption that will ultimately lead to the purest of vital health. Find your secret formula and then you are off like a rocket.

Your personal alchemy at any given moment is energetically holding you in a place. You can only interpret reality from this one place, this one level of consciousness. The alchemy you consume creates your vibration. It generates your quality of life. While there, you can only experience things at that vibrational level. You cannot perceive or be aware of any other level of consciousness. The trick is to know there is always somewhere more to go.

Things in your life either evolve your vibration, or de-evolve your vibration. You get to figure out which one you are going to choose. It is wise to be responsible for your vibration with education about what you are consuming. Learn to refine your personal alchemy, so it is beneficial for you. Life has rhythm. You can figure yours out, thereby enhancing your personal flow.

A wise place to start in discovering your personal secret alchemical formula is by analysing how your alchemical intake makes you feel. In the moment of consumption, your body will always give you indicators of what is going on with it. You just have to listen.

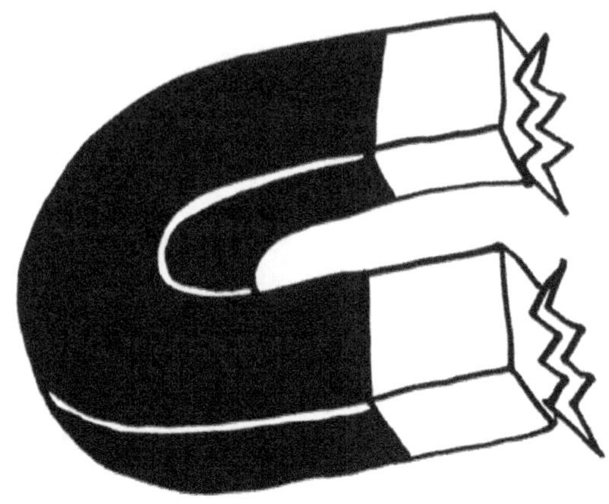

# Earth magnetism

The Earth is magnetic. All of the creatures of Earth are magnetic. They all have a magnetic field that is the source of their power. It holds them together. You are an energetic and vital creature. You have a magnetic field. You need to ingest magnetism in order to maintain and grow your own magnetism.

Your personal magnetism is your power. It is your sparkle. It is the source of your vitality. It is your ability to experience bliss. It is your life force. The more magnetism you have, the more soul you are able to express. Having low levels of magnetism makes you lifeless. Living becomes difficult because you are closer to dead than alive. Magnetism is the battery of the world. Without it there would be no life.

The land is intimately connected to food production. Food imprints magnetism from the land. Then you ingest the magnetism from the food. Vital health comes about by digesting potent minerals and nutrients empowered with the magnetism from the land.[7] Their magnetism empowers your own magnetism.

More than just nutrients, vitamins and minerals, the body feeds off the magnetism given to you by the Earth herself. Magnetism is present in food. The intensity of the vital energy of its magnetism or life force depends on where the food has come from, who harvested it, how long it has been harvested for and the degree the food has been processed.

Personal magnetism comes from the potent magnetism that you consume. What you are looking for is empowered minerals. Minerals are more than nutrients. Traditionally, humans had an intense connection to the land. This connection is the channel for the magnetism to flow from the Earth into you. It gives humans a natural, constant source of energetic fuel.

Food is being interfered with so much today, it is quickly losing its magnetism. This is affecting your health so drastically it is funny. Food that has been manufactured, chemically altered, made in machines, genetically modified, microwaved or irradiated has lost most of its vital energy or life force. It is practically useless as far as your body and vital health are concerned. Soulless food equals soulless humans.[8] It is wise to throw out your microwave right now.

It is wise to avoid vibrationally dead foods. The body still has to process it but gets little nutritional value from it, thereby taking your vitality instead of giving you vitality. This is detrimental to your health in the end because it robs you of your sparkle.

Magnetism is life force. It is the vital energy of being alive. It is the sparkle of life. Without it, life cannot flourish and grow. Magnetism is the true currency of Earth. It is wise to bring awareness to your personal magnetism and learn how to cultivate it.

# real food

Real foods grow naturally in Nature. They have not been tampered with by humans. Whole, natural food includes vegetables, meat, eggs, grains, fruit, nuts, seeds, herbs, legumes, spices, and seafood.

Eating organic food leads to vibrant living. No wonder organic food sales are growing at a tremendous rate. Organic food tastes better, is better for you and has more nutrients. It supports agriculture that is beneficial for the Earth and all her creatures. It empowers the soil instead of killing it with harsh chemicals. You know what you are eating because organic farmers are honest about their food production. It is quite difficult for them to be certified organic.

It is wise to be skeptical of labels that say "organics" and the like. They can be companies that are trying to scam you. You can learn to read labels, to know what the ingredients actually are. You can educate yourself with some research.

Eating local is smart. Support your local farmers because you have a much more powerful magnetic link to food grown near you. Therefore, it can increase your own personal magnetism. Eat what is in season. Focusing on local food production empowers your local economy.

Growing a garden is a great way to be connected to your food and the Earth. It is a firsthand experience with Nature. It gets you outside and working hands-on with plants. This can directly increase your magnetism.

It is better to drink water that is alive, rather than bottled water, which is dead. Water can be purified by flowing in a spiral. You can use two magnets on your water pipe to make it do so.

Real food contains fiber and this fiber is the best way to maintain your personal flow. It is wise to include antioxidants, magnesium, enzymes, green supplements, fiber supplements, fish oil, probiotics, blue-green algae, and enough protein into your diet in order to be a strong and healthy human.

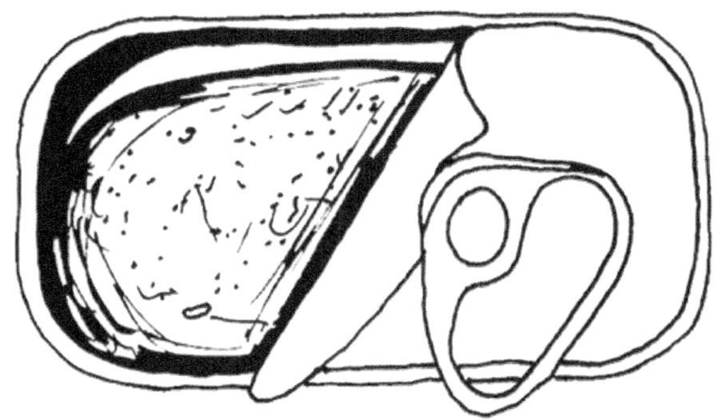

## processed food

Processed foods are foods with additives or sweeteners or that have been canned, dehydrated, exposed to aseptic processing, frozen, refrigerated, bottled, dried, pasteurized, concentrated, fried, stabilized, turned into powder, or preserved. Overeating junk food makes and keeps you fat and unhealthy.

This includes crackers, cereal, bread, pasta, bakery goods, chips, frozen foods, salad dressings, sauces, boxed mixes, packaged anything, processed meats, candy, canned goods, fast food, precooked foods, things in jars, drinks, desserts, and chocolate, just to name a few.

Processed foods have lost much of their nutritional content. They have hidden ingredients. They use genetically modified foods. They can be reconstituted with synthetics. The repercussions of the additives are unknown. They have been exposed to metal contamination and metal lubricants from the machinery that processes them.

Food that is not organic is host to many chemicals such as fertilizers, hormones, artificial stimulants, fabricated chemicals, preservatives, antibiotics, artificial flavour, synthetic dye, excess salt, herbicides and pesticides. These chemicals cause cancer.

These chemicals affect your brain, body and immune system in negative ways. They do not assist you in your personal evolution. These harsh chemicals are turning up in the ground water, in fetuses and in your blood. They are totally affecting the food chain in unknowable ways. It is not wise to mess with such toxicity.

Non-organic food can be irradiated, which means x-rayed with radiation to kill bugs. Radiation is nuclear, which is radioactive. Eating food exposed to radioactive chemicals kills all life force. It is not wise to eat dead food because it has little benefit for the body.

The reality is that genetically modified foods are a real threat to humanity. By tampering with genes of food, scientists are splicing the life force out of it. Food without life force is useless for humans. It creates humans without life force. It is wise to stick with food the Creatrix has created for you, instead of Frankenfood.

Refined products have been processed to the point where their magnetism or vital life force, nutrients and minerals have been stripped away. They leave food not only nutritionally valueless, but taxing on the system. Refined sugar and salt have been harshly processed and as a result are so intense as to be poisonous for the body. It is wise to be in control of your intake of these toxic things.

Refined products rage through your cell walls, causing much damage. This directly results in aging of the body. Processed foods tend to have a high percentage of sodium, which makes them quite lethal for the heart.

Refined products have been processed so much it renders them addictive. They are concentrated forms of what your body needs for survival, so of course your body wants them. But, they have been corrupted and no longer serve the body well.

There are natural sugars that are way less harmful for the body like honey, natural palm sugar, stevia, xylitol, agave nectar, molasses, and even organic sugar, which has retained some of the nutrients and minerals, as some examples. It is wise to avoid sugar altogether as it is very acidic.

There are also more natural salt alternatives. Sea salt is better, but it has still been processed. You can buy Himalayan rock salt. It is the least refined salt, so it contains many of its natural nutrients and minerals. This makes it a wise choice.

A whole food diet is simple. Humans are sadly so far from collectively eating properly and maintaining vital health. You have chemical and elemental addictions to the things you consume. A spell of processing has been placed upon the elements that makes you addicted to the actual process itself, not the food. Choosing artificial food leads to being desensitized and numb from having to process the artificial elements within it. This is contrary to vitality.

Food is broken down into simple molecules in your body, and then you absorb them. Unfortunately, some of the elements that have been processed have irreversible corruption done to them. This follows the Law of Microscopic Reversibility, a Law of Nature that says that all chemical reactions can be reversed except for those that have an energy bias. This fuckedupness causes your body damage when you ingest such foods. They can never be broken down properly into the simple molecules that the body needs to function.

The food you take in is a reflection and the foundation of your consciousness. When you eat processed foods, you are not functioning at full power, even though you may not realize it. Because you are inundated with toxic shit soup, you may not be aware that it is making you act and do stupid things. But, you really can be.

eating animals is sacred

Animal meat is the highest vibrational food you can eat. It is the closest in vibration to you. It has the highest content of magnetism and life force for you to consume. The spirit of the animal is in the meat. When you eat it, the animal learns there is a higher vibration for it to achieve. You are helping the animal on its reincarnational path by consuming it.

You know when you meet a person who you think is super cool, and you want to be like them? Well the same goes for the cow! He gets excited and stimulated by the intense energy rush he experiences inside of you.

The cow wants to emulate your energy and raise his own vibration. Now he knows there is somewhere more to go. You are opening the door and showing him the staircase. This assists him to reincarnate at a different energetic level the next time round on the wheel of life.[9] This is how the world turns.

By eating animals, you stay intimately connected with the Earth. It is something humans have been doing for millions of years. It is your natural way of being. It flows with the Laws of Nature. Animal fat is how humans have managed to survive for this long.

Eating animals is sacred. If there is a problem with eating animals, it is the way they are being killed for the supper table. There is no reason why society cannot evolve to having more respect in the slaughtering of animals. Overpopulation is the real issue concerning animal consumption.

It is wise to buy meat that is killed in a sacred and humane manner. There are slaughterhouses that do just that. There are organic farms that are smaller and more harmonious in their treatment of animals and their meat. The belief that eating animals is wrong and eating synthetic processed foods, like supplements, is right, is a concept that is out of balance and fatally flawed. It does not flow with the Laws of Nature. An issue with eating animals is overconsumption. Meat has a high content of acidity. It is wise to balance it out with vegetables.

When you go to the grocery store, are hunting, are preparing food, feeding your pets or just about to eat an animal, it is wise to open your heart and with intention say, "Thank you for dying so I may live." Animals want their deaths to be acknowledged and gratitude to be shown. This makes their death worthwhile, and not a waste. The vibration of gratitude extends into the present, past and future and wraps the animals in love. This energy supports and nourishes animals.

Ignoring animals by not eating them does not assist them at all. In fact, it hurts them deeply. They feel you do not care because you do not participate with them.  Nor does it affect change at the heart of the matter, which is the way they are raised and slaughtered.

Plants feel just as much as animals do.  When they are killed they are sensitive to it.[10]  When you cut plants down, they grow back even more. The plants want to be cut down because that is how they are in service to the Creatrix.  It is the same with animals.  They want to be eaten because it is their job.  They have been doing it for the entire life of the planet because Nature thrives on consumption.

The entire Earth functions upon its creatures killing and eating each other.  Every single species, every animal does this.  Minerals willingly die for plants.  Plants willingly die for animals.  Animals willingly die for other animals higher on the food chain.  Consumption is a foundational Law of Nature.  To deny it is painfully ridiculous.

Animals do not mind being killed because they reincarnate.  They never end and they inherently know this.  They do not fear death like humans do.  It is never an individual situation.  It is only your limited perspective that makes it anything else.  Death is ecstasy.

Animals have a group soul that absorbs pain and death.  If one is experiencing trauma, he is immediately surrounded by billions of others who absorb what he is experiencing.  They break the pain down into such smaller units that they share, that there is no suffering.  They are able to take their individual presence and get lost into the collective.  The pain goes through them.  But this does not mean that they do not feel scared, uncomfortable or frightened.  They are incredibly sensitive.

For example, if a plant is being cut down, it takes itself and moves its consciousness into another plant, of the same kind, somewhere else. This leaves the plant without feelers so it does not have to suffer.  The morphogenetic field allows the spirits or consciousness of plants and animals to travel within the collective because they are completely connected to one another.  There is no separation amongst them.  Humans are the ones who are separate.

There are terrible atrocities happening to animals all over the world. They are being hunted and killed without honour.  In the first world, you sit back and cry out in horror over the practices of those less fortunate.  But, put yourself in their shoes.  Many people of the world are starving.  Starving. When a human is starving and has many children to feed they are willing to do things well-fed people would not do in order to survive.

When you are hungry the rules for life change.  So, no wonder people are hunting and killing animals in ways that you in the first world find abhorrent. They are so hungry they do not have the time or capability to care like others do. It is not right for them to be judged. If it came down to it, odds are, you too would desperately kill in order to survive.

Before there used to be small farms that raised animals for food. It was farmers and their families who loved and took care of the animals. When the animals were slaughtered that loving relationship was consummated.

With the invention of massive farms and industrial farming, out went the priorities of actually tending and loving the animals. The relationship between people and animals became distant and cold. No longer did the animals feel loved. Now the slaughtering of animals is impersonal. The animals are afraid. You are now consuming this fear and cold, uncaring behaviour.

The industrial meat industry does not have the best interests of animals at heart. They are scared and alone. They are slaughtered without love. This is the real tragedy of industrial farming. All for the profit for the few. It would be wise for humans to revert back to the old ways of animal husbandry by reinstating small farms.

## wheat sugar dairy

Wheat, sugar and dairy are the main staples of a modern diet. The only problem with these things is that the human body is not designed to digest these high-intensity foods to the degree that they are now being consumed.

The result of this overconsumption is discomfort, food allergies, and ultimately sickness. You see this happening more and more. Humans have only been practicing agriculture for roughly 10, 000 years. Wheat, sugar and dairy have not been in your diet for long.  You have not had the time to adapt to them.

Humans are not able to digest fully the protein in wheat.  It has been genetically altered over many years to have more protein.  This makes it even more difficult to digest.  In effect, it is poisoning humans due to overconsumption.

Wheat does not have any nutritional value.  In fact, it contains things that are unhealthy, rendering it useless in the diet.  It is wise to avoid it.  Flour has no nutritional value.  It is glue and it slows down your digestion to a snail's pace, robbing you of your vital health.  It is wise to avoid the monocrops such as wheat, corn, soy, and rice.  It is better to enjoy less fucked with grains such as millet, buckwheat, amaranth and quinoa.

Refined sugar is a powerful drug.  It affects emotions and mood big time.  It is so difficult for the body to process, that it has to take minerals out of the bone marrow to do so.  This imbalance can eventually cause aging and sickness.

Sugar is so caloric that the body has no way of processing it.  It stores almost immediately as fat, making it difficult for your body to process this fat.  It slows down your digestion as well.  Food high in sugar turns your flow into a trickle.

Sugar is highly addictive. It is poisonous for the body.  It is responsible for disease and death. Cocaine is illegal for these reasons.  Sugar and cocaine have similar effects upon the brain.  So why is sugar dismissed as harmless?  Why are all of the holidays centred on this powerful drug?

Children should not have sugar at all.  It affects their little bodies the same as if you had given them crack.  It sets up a cycle of addiction that will be a problem for them the rest of their lives.  Just because practically everyone on the planet is addicted to sugar, does not make it right.

There is an epidemic of children with major addictions to this drug that is ignored, and downplayed by society.  It is no coincidence that there are problems like attention deficient and hyperactivity in children because parents let them consume too much sugar.

Non-organic cows are pumped with chemicals such as antibiotics and growth hormones.  You consume them directly when you eat dairy that is not organic.  The fat content of dairy is too high for your body to consume regularly.  This fat goes into your intestines and seals them off.  This leaves your intestines unable to process nutrients, leaving your body starved for them.  You know this because of how difficult it is to clean cheese from a dirty plate.

Dairy is highly mucus-producing. This mucus is a perfect breeding ground for bacteria, fungus and toxicity. Your body just cannot handle that much fat. It causes much obesity, due to overconsumption.

Dairy, especially cheese, slows down digestion big time. Having a slow digestion makes you fat, harbours toxins and pathogens, affects immunity and slows down your body's internal processes. This leads ultimately to poor health, a shorter life span and a rotten quality of life.

Wheat, sugar, and dairy are causing an epidemic of obesity and food allergies, as people's systems protest. These allergies cause the body to become inflamed. Your emotions get tied in with the agitated feelings your body is giving off. But, it is not you. It is the body sending out a distress signal.

The symptoms for food allergies are similar to environmental allergies; you get the same symptoms. Most people may not realize they are allergic to the food they are consuming. It is not wise for humans to be eating food that is not digestible for their systems.

# clothing

Wearing clothes made from organic fibers like silk, cotton, leather, bamboo, linen, wool and hemp is a way to go for ultimate vibrational experience. You can buy organic clothing! Your body needs to breathe. Wearing natural clothing helps your body function properly. Pure fibers can raise your vibration. They aid in your ability to express and move energy.

Non-organic fibers have been made with harsh chemicals in the manufacturing process. These chemicals are present in the fabric. You are exposed to them when you wear the clothes. Avoiding them is wise, since synthetic clothes make it difficult for the body to breathe. The artificial fibers literally seal you in. They can halt your personal flow of sweat, energy and movement.

Polyester is a shallow vibrational fabric, as it is plastic. The plastic of the polyester releases its chemicals when warmed by the skin. This constantly barrages you with toxicity because the skin absorbs these chemicals. Especially avoid when exercising, as you are likely to be breathing deeply. Synthetic fibers include polyester, nylon, acrylic, polyolefin, and spandex.

The clothes industry tries to sell rayon and viscose as natural fibers. But, they are not. They are processed, though their source is wood fiber. They are considered semi-synthetic fibers. Usually synthetic chemicals have been added for strength and durability. It is wise instead to have things that are of the Earth next to your skin. You are what you wear.

Have you ever contemplated making your own clothes? Or maybe getting clothes made for you? Clothes made with love are the highest vibrational things you could wear.

# your hair

Your hair is important because it is your personal antennae. It enables you to be vibrationally sensitive and connected to the Creatrix. It can empower your ability to hold the lightning of bliss. Your hair is your awareness. Cutting it takes away your ability to perceive and to know. Your hair is your wings. It is sensitive. It is important to protect it and take care of it.

When you dye your hair, you are killing this sensitivity. This weakens your connection to the Creatrix. It dulls your ability to feel. It is wise to let your hair be its natural self, by not dumping tons of chemicals into it.

Washing your hair with shampoo and conditioner every day strips it of its natural ability to maintain itself.  The corporations who are selling you these products make it so your hair becomes addicted to them, so you buy more.  By robbing your hair of its natural luster and then replacing it with a chemical luster, you are not doing yourself any favours.

Your hair does not need to be washed as much as that or by such harsh chemicals.  Natural shampoos and castile soap are mild yet effective for your hair.  When you need to condition it, coconut oil, jojoba oil, argan oil, cocoa butter and natural conditioners all work wonders.

It is wise to avoid putting in your hair plastic, metal, hair dryers, useless chemicals, and curling irons, as over time these are highly destructive to your delicate hair.

# living free of chemicals

Chemicals are not normal.  Their energy is chaotic because it has been fabricated and not birthed into the world via the Creatrix.  There is no way to measure the negative effect chemicals have on your life or the life of the planet.  The negative consequences being created by having so many chemicals in the food, products, soil, animals, air and circulating on the Earth are becoming more clearly evident.

Science is now discovering chemicals in places they should not be, like mother's breast milk, groundwater, bone marrow, in baby animals and in the ocean.  Common sense says the more the world is poisoned, the more difficult life will become.

Humans have no idea of the consequences of using chemicals to the degree that they do.  Chemicals are not natural.  The Creatrix did not make them.  They cannot be trusted.  Chemicals take your unlimited potential and greatly limit it.

It is not so much the molecules of chemicals that are the problem.  The Law of Microscopic Reversibility, a Law of Nature, says that most molecules can return to their natural state. It is the actual process of making the chemicals that flies against the Creatrix and makes them dangerous. Some molecules are forever changed. They are unable to be broken down into simple molecules, as they should be. Therefore, they are unstable and unpredictable.

There is a monumental difference between something birthed by the Creatrix and something fabricated by a human or machine. It is the intention behind the processes that are the concern because they are not based on the best interests of humans. Whereas the Creatrix always has your best interest at heart.

Every chemical you touch and come in contact with goes inside and through you. Exposure to chemicals can have a detrimental effect on your body, your brain, your blood, your liver and kidneys. They have to work hard on cleaning up after exposure. They tax your system's resources. They are difficult for the body to process. They can cause discomfort, aging and eventually disease.

Digesting chemicals greatly lowers your vibration because they are so hard on the body, especially with overconsumption. They take away the body's ability to heal itself because they suppress your immune system. Your ability to have free will to decide on your exposure level is being taken away because harsh chemicals are now present in the air you breathe.

Choosing biodegradable, non-irritating, non-poisonous, phosphate-free soaps and cleaners is healthy for the environment and for you. Otherwise, they can be poisonous for the land and wildlife. They go into the ground water and are toxic to the water and wildlife. Toxic means they kill the creatures of the Earth upon contact.

Why would you want to use products that are directly killing the planet and her creatures? You can switch from bleach to hydrogen peroxide. The thing with clean is that it has no smell. Something is clean when it has a lack of smell. How did you get convinced you needed a slew of artificial fragrances for something to be clean?

Using toilet paper and feminine products that do not use chlorine bleach is a perceptive way to avoid chemicals in your sacred parts. Chlorine bleach is poisonous to humans. You can buy products made with hydrogen peroxide instead. They are much less taxing on the environment and your body.

Synthetic fragrances are neurotoxic! Perfume, cologne, aftershave, deodorant, body soaps, hair products, cleaners, laundry soap, facial and body creams, makeup, shampoo and conditioners, and air fresheners contain artificial chemicals that are poisonous to humans. With prolonged exposure, they reduce your oxygen intake and retard brain functions.

Synthetic fragrance coats your lungs, veins and brain with chemicals that are practically plastic. They make it difficult to breathe. The oxygen has to fight to get into your delicate passageways. Oxygen is the number one requirement of the body. It is wise to make it a priority.

Artificial fragrance can make you lose your ability to smell yourself. But, the rest of us can certainly smell you. One application can last longer than a week. It is a strange phenomenon when you cannot breathe properly because someone near you is wearing synthetic perfume. It fills the entire room with its insidious poison, affecting the oxygen intake for dozens of people.

It is barfy when you walk outdoors at night and all you can smell is some neighbour's overpowering laundry soap. It is an invasion of privacy to be exposed to such stinky, harsh chemicals when you do not want to be.

Essential oils and aromatherapy work great for your fragrance needs. Pure smells are often a forgotten part of life. Essential oils are real. They come from natural sources. They are usually not contaminated with chemicals. Buying organic is wise.

Essential oils can play an important part to a happily functioning brain. They make for great perfumes. You can put them on your clothes or in your hair. They can irritate the skin with contact because they are quite concentrated.

It is wise to have a special pot of water on the stove on low and add a few drops of a few different essential oils. This creates a wonderful fragrance that will fill your home with natural aromas. You will be surprised at all the various beautiful smells you can create when you combine several oils.

The aluminum in deodorant is poisonous. It goes inside of you and the metals stay in your blood. They can eventually collect in the brain. Metals in the brain cause Alzheimer's. It is a wise idea to get a salt deodorant crystal instead. They work great.

Pharmaceutical drugs are synthetic and interfere with your natural processes of life. They greatly affect your ability to heal yourself. They can suppress the immune system. They usually mask the symptoms but do not deal with the source of the problem. Your body can get sick because your behaviour is not right for you. Change your behaviour. Change your susceptibility to disease. Therefore, change your need for pharmaceutical drugs.

Pharmaceutical drugs are made from artificial materials. These synthetic chemicals get in your blood. It is difficult for your body to process them. They are taxing for your system. Pharmaceutical drugs help you by hurting you in other ways.

If you have a headache, it is because your body is trying to send you a message that your behaviour is not working. In order to be free of it, you must figure out the lesson it is teaching. Suffer and go through it, change your behaviour and it will not happen again. Take the painkillers and you will get another headache because you have not figured out what the problem is.

Odds are the headache happened because you did not drink enough water. Your organs did not have enough hydration to deal with the chemicals you consumed that day. Water is the second most vital requirement of the body. Sex and drinking water are the best cures for headaches.

Why take drugs to mask natural reactions like jetlag, headaches, nausea or eye irritation? The body is trying to communicate with you through these natural phenomena. There is no harm in going through them. The idea is to have a functioning relationship with the body. It is wise to listen to it and respond. Ignoring its messages and overriding them with synthetics greatly lowers your personal vibration.

Additives in foods like artificial flavouring and dyes are not healthy for human consumption because they are artificial chemicals. They are inorganic. They are plastic. They are poisonous.

Synthetics are used in making large-scale, mainstream beer varieties. They go into your body when you consume it. There are healthier, smaller brewery choices. If you have to drink, it is wise to try organic varieties of wines and other forms of alcohol. They will reduce the level of headache you have the next morning.

Hydrogenated oils including margarine, canola oil and corn oils, have been heavily processed for a longer shelf life. This turns them into something equivalent to indigestible plastic. This is why overeating fast food like fried foods, baked goods, crackers and potato chips has negative effects upon the human body.

Hydrogenated oils, also contain artificial trans-fats, which are lethal for the heart. There is absolutely no reason to use these kinds of fabricated chemicals in food other than to increase shelf life in order to increase corporate profits.  They are a major cause of heart disease and death in the population.

MSG is poison.  Sodium nitrates found in meats cause cancer.  There are meats, like bacon, available without sodium nitrate.  Foods in metal cans can leak metal toxicity into the food.  Farmed fish are not swimming as they should.  Therefore, they are a nest for bacteria and toxins.  If you are buying fish from your grocery store odds are it is farmed fish no matter what the label says.

Aspartame, artificial sweeteners and fluoride are poisonous and found in lots of places like gum, soda pop, candy and toothpaste.  Just because they are there does not mean they are cool.  All these harsh toxins affect the brain and body in ways that totally suck.   Exposure to too many chemicals leads to cancer.  It may not seem like a big deal in the moment, but a lifetimes of uneducated choices leads to sickness and death.

## petroleum products

The major source of the artificial chemicals used in food and personal products is crude oil, coal tar, coal or natural gas.  These petroleum products consist of petrochemicals.  They have the same ingredients that are in gasoline, plastic, lubricants, propane, and rubber, as a few examples.

Digesting petrochemicals does not lead to vitality.  It depresses vital health in humans.  Check your ingredient lists because artificial flavouring is in most processed foods.  It is wise to ask yourself why you are eating crude oil.

Toothpaste, eyeliner, nail polish, lip gloss, sunscreen, lipstick, foundation, hair mousse, hair spray, lotions, hair gel, hair dye, mascara, body and face moisturizers, cleansers, perfume, gum, sunscreen and most other personal products are made of petroleum products and many different kinds of toxic elements. This means you are putting on your face and in your belly the same thing you fuel your car with. This is far from logical.

Petroleum products contain harsh petrochemicals, heavy metals, and toxins. They are completely indigestible by the human body. These toxic chemicals go into the body and in the brain and other organs, like the liver and kidneys.

This kind of chemical buildup leads to Alzheimer's, heart disease, strokes and cancer. These products tax and stress the immune system. They totally disrupt your body's ability to function properly. They are poisonous for humans.

Petroleum products change your alchemy and make you uncomfortable because of the effect they have on your pH levels. This discomfort is felt at a low level of awareness, and can make you constantly irritated. This irritability puts a cap on your ability to feel bliss.

Because you are uncomfortable, it does not allow you to feel comfort. Ingesting petrochemicals makes your body work excessively hard to rid itself of the toxic overload of shit. In essence, the body goes numb and shuts down its ability to feel normally. It is wise to ask why you would want to desensitize yourself.

However, it is a choice. You can commit to not buying petrochemicals, and make an effort to go and find new products to buy, because there are tons of alternatives to petrochemicals. There are natural alternatives to every single product out there. You just have to decide to be a normal human instead of a synthetic human. When you put something on your body, it is the same as eating it. It enters the bloodstream in mere seconds.

It is wise to make sure the ingredients are natural. Many products will explain exactly what the ingredients are, so you can make informed decisions about what you are consuming. It is wise to be wary, though, because some products do everything to trick you.

Most beauty products are made of petroleum products. They are poisonous and taxing for the body to process. They cause aging, the opposite of the beautifying qualities they claim to have.

You may not realize that the artificial molecules go out from the artificial fragrance you use and into your living environment. They coat everything they come in contact with. They create layers and layers of crap that your body must process all the time. This is a real drain on your immune system because you constantly have to deal with high levels of chemicals in your system.

Maybe using just a few products a day, your body can handle. But, when you use 93 different chemicals to get ready in the morning, from your clothes, to your lipstick, to your breakfast cereal, your body quickly can become overloaded with toxicity. You keep dumping tons of chemicals into your system, and eventually it makes for a poor vibrational existence. Your spirit is squeezed out of your body by all the toxic sludge that resides in your blood.

## toxicity

All of this exposure to chemicals adds up and you become toxic. The toxicity resides in the fat on your body, in your organs like your brain, liver and kidneys, and in the fat around your heart. This makes normal functioning quite laborious for them. Toxins suck!

When you are toxic, your personal environment is energetically lethal. You will get sick. It is just a matter of time. It affects the quality of your life to the point where concepts like spirit and soul are like TV commercials. After a while they do not register.

Your liver and kidneys are your toxin filters. They can get overwhelmed from a lifetime of chemical choices that de-evolve you. Your lungs also get greatly affected by these harsh chemicals, as you are breathing them in constantly.

These organs are vital. It does not take much to damage them. Once they are hurt, it makes it not easy to come back from it. When you are functioning with self-damaged organs, it makes it difficult to maintain a healthy energetic system.

All humans have toxins stored in the body. The world is inundated with pollution and harsh chemicals. They are in the air you breathe, the food you eat, the clothes you wear, in the drinking water. You are now born with them in your body. Even the healthiest of people still have traces of toxins present in their blood. These toxins accumulate over your lifetime.

Any fat you have on your body stores toxins. Fat is unused fuel. Your emotions tend to fester in the fat. For example, when you are depressed and you do not like yourself, you go and buy synthetic fast food, full of poisons and toxins.

When you are eating the food laced with synthetics, on some level you can be thinking, "I am unhappy, I do not like myself." If you cared about yourself, you would not be eating food that was poisonous for the body. Your body is unable to use the overabundance of fuel you are giving it, so it stores it as fat. These emotions get caught inside the fat you are making in the moment.

You then have layers and layers of fat. These are layers of emotions about how much you do not like yourself, all for everyone to see. Mingling in your fat are the toxins you digested as well. This toxic fat soup is like a huge bubble suit that you wear. You cannot breathe; you can hardly move, let alone function normally.

The outcome is intense agitation that you radiate outwards, that other people can easily pick up on. You are grumpy and have toxic emotions. You can realize it is not you but the fat that is toxic. You can work at getting rid of the fat, and you will notice you no longer have toxic emotions.

As a human, one of the things you can commit to is being free of synthetics as much as possible. You can do this by taking responsibility for your blood. Your sparkle power resides in your blood. It is one of your most important assets. Your blood is the home of your spirit.

Everything you put on or in you ends up in your blood. Your skin absorbs all of the products you put on it. The chemicals then enter the bloodstream. If your veins have a bunch of artificial and toxic particles in them, it is difficult for your body to function properly.

The synthetic wages war upon the sparkle that resides in your blood. It brings disease and death. When you have more artificial than spirit in your veins, it makes you not human. It pushes and squishes your spirit down and out. Therefore, you have made a choice to not be a real human. Not to be guided by spirit, soul or the Creatrix. Instead, you have allowed the synthetic to be your guide.

Your experience of your future is in your blood. Your alchemical choices dictate your reality for many days ahead, even longer if they are artificial. It takes the body that long to process whatever you have consumed, whether it is by being absorbed through the skin, breathing or eating.

What you consume and are exposed to accounts for your emotions, your brain functions, your personality, your mood, your ability to be intelligent, your ability to think, your actions, your ability to feel, your way of being, your level of consciousness and everything else that is life. It is rather important to be aware of the consequences of your alchemical choices.

Anxiety is created by foreign particulates in the blood. Your system reacts negatively to being invaded. It produces anxiety because it is communicating to you that something is out of balance. Your body and feelings are always communicating to you about your alchemical state because they are trying to work with you. They create anxiety in an effort to make you understand that what you are consuming is not for you. It is wise to listen to them instead of ignoring them. Not listening to their warnings desensitizes you and makes you numb.

Particulates in the blood make you fake. When you consume inorganic alchemy, it pushes your "I am" presence to the side. You are then not yourself. Therefore, you become fake. A controllable slave, with little power or free will. Instead, you do what you are told, what the contents of your blood are dictating to you. Unfortunately, synthetics in the blood can set up a cycle of addiction that you easily fall prey to.

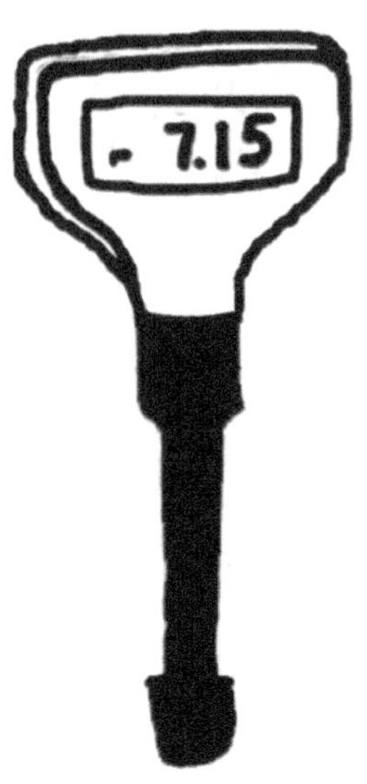


## pH

Your blood maintains itself via its pH. A healthy human has a balance between alkaline and acidic. This is a precise balance, leaning towards alkaline more than acidic. It is the starting point for your vital health. Its balance is a reflection of being centred. Your pH is the lubrication of your liquids as they move through your matter. Your pH grows either magnetism or parasites depending upon its balance.

Your pH is what creates the surface tension of your cell membranes. It also creates the surface tension of your aura. Obviously it is what is holding you together. If it is too acidic, holes and weaknesses start to appear. This is when the trouble begins. It leads to disease and death. Simple.

What you consume goes into your blood and creates your pH. This directly creates your emotions, state of mind, place of being, your ability to communicate honestly with others, and your consciousness. You can be grounded, be centred, have energy, but if your pH is out of balance, then you are off.

Your pH directly affects your blood. Your blood is the house of your spirit. Your pH is the direct communication between your body and your spirit. Your pH also communicates directly with your organs. The importance of the balance of your pH cannot be overstated.

Foods, including processed foods, dairy, meat, junk food, fried foods and sugar all are acidic. Culturally humans consume way too much of them. When the blood becomes too acidic, it is a perfect breeding ground for bacteria, mucus, fungus, and parasites. This kind of imbalance in your blood will eventually lead to disease.

Alkaline foods include fresh lemons, avocados, green veggies, stevia, and apple cider vinegar. These types of food promote clean, bacteria-free blood, as well as healthy flow of the bowels. Focusing on fresh whole foods instead of processed foods is the way to maintain your alkalinity.

The pH of your blood directly causes your thought patterns. Any imbalance in your nutrient uptake causes your pH to be off. This is how obsessive thought patterns begin. These obsessive thoughts are the opposite of the flow of life. They are restrictive and confining.

You are a tube. You are a complex worm. Your pH lubricates this tube from your mouth to your bum. Make that connection by feeling what you eat and how it affects your shit the next morning, because they are vitally connected. Junk food is acidic glue. Fiber is alkaline lubrication.

The secret to life is alchemical. Bliss is alchemical. It starts with your pH. Society will try to poison you with its collective ignorance. You will poison yourself if you are not aware of the consequences of your alchemical actions. Having a little poison is the same as having a lot because it throws your pH off.

Refined salt and sugar, chemicals, and toxins have a detrimental effect on your cell walls. They are abrasive. If your cell walls are damaged, it directly results in aging and sickness. Particulates in blood create friction, heat and aging which directly fuck with your cell walls. Your pH and ultimately your flow is dependent upon your cell wall permeability being undamaged.

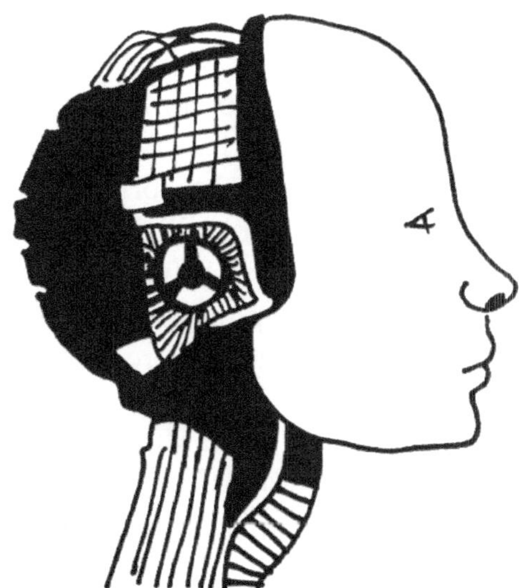

synthetic sickness

There is a monumental difference between putting refined sugar in the hummingbird feeder and planting flowers that the hummingbird will suckle from. You are surrounded by inorganic and synthetic things. They are not natural or normal and they do not promote vital health. Synthetic is meant to be a tool, not a dictator.

The problem with synthetics is when they get inside of you. They wrap you in a plastic coating. If your organs, clothes, food and environment are plastic, it seals you off from the real, organic world, into a little pseudo pod. It gets into your brain, blood and body. It puts artificial barriers between things where there should be a natural flow happening. Cyborgs are synthetic. Humans are not suppose to be synthetic.

There is only one reason why society turned from using real ingredients in its alchemy to using artificial ingredients, and that is corporate profit. Corporations make huge profits from using non-organic ingredients. They are not concerned with the well-being of humans at all. Your vital health does not create as much profit as your sickness does.

When humans are on Earth, the Creatrix knows that they need food and shelter. When you are so ungrounded, so disconnected from them, that the Creatrix does not know you are there, and does not know to provide for you. Living an artificial life is like living in a condom. You cannot feel and the Creatrix cannot feel you.

Being this cut-off from Nature is an inorganic way to live. It affects the way you think, the way you feel, your behaviour and the way you function within the human collective. It lends to a robotic, clinical and unnatural way of being. So many people of today are dislocated from the Creatrix. The Creatrix does not know they exist.

If you live 20 stories off the ground, you wear rubber-soled shoes, you walk on concrete, you get in your car, you are in the office all day, and then go back to 20 stories off the ground at the end of the day, you eat boxed food, full of artificial flavouring, you wear synthetic clothes, you are covered in artificial fragrance from your deodorant, shampoo, perfume, laundry soap and plugged-in air freshener, your veins are full of synthetic pharmaceutical drugs, petrochemicals, alcohol and nicotine, and you rarely touch the Earth, the trees or the animals, then you do not have much of a relationship with the real world.

Your sparkle can be near dead or gone altogether. Sparkleless people are not human. Sparkleless people do not have a connection to their souls. It is wise to ask yourself, then what the hell are they? Who or what is driving their bus?

There is a difference between walking on the concrete (straight and hard) and walking on the Earth (curvaceous and uneven ground). You need that uneven road because your brain is uneven. Concrete makes your brain hard and flat because you imprint your environment.

There is a difference between being a real, sparkly, vital human and being a fake, automatic, plastic human. Are you a real human or slowly becoming a cyborg human? Why risk yourself needlessly by constantly consuming artificial alchemy? Question what you are purchasing, especially if it is synthetic. It is wise to ask yourself if this artificial product is going to make you more of who you are. The answer is no.

Fake emotions are a result of fake alchemy. Being fake is having that tone of voice that goes up high at the end of a sentence. It is telling people what they want to hear, what is expected of you or what etiquette dictates, instead of expressing how you really feel. You do not have to participate in the fakeness. You can just step aside. If you want to take the red pill that will show to you the true nature of reality, then stop consuming alchemical synthetics.

Any kind of exposure to toxicity immediately causes the systems of the body to react in a negative way. Somehow, poisons have become acceptable forms of personal fuel in society. That they are even considered okay to digest is a crime. They are just poisonous, artificial drugs. They are synthetic because they are processed and they are energetically dead.

How did society get convinced it was logical to consume dead alchemy without thinking it was destructive? Synthetics make you not care. They make you unemotional, aloof, callous, detached, cold, passionless and heartless. This not-caring is a sickness. When you are not connected to your heart, your spirit and your soul, you are not a real human. If you are not a real human, it is wise to ask, what are you? You certainly become a hardworking employee.

The synthetics go inside of you, coating you with plastic. This plastic seals you in and makes you numb. The numbness makes you not care about yourself or the harm you are doing to yourself. It leads to not caring about others and the world around you. This uncaringness is an epidemic. The Earth is dying because of artificial people who do not care.

Humans are addicted to white powder as a society. Refined white powder is poison. Simple. Refined sugar and salt, pharmaceutical drugs, caffeine, MSG, artificial flavouring and flour are all white powders that are deadly for humans.  They are severely potent forms of fuel.  They damage the body because of their intensity.

Things that have been refined wear out the cellular membranes with their harshness.  They are so difficult for the body to process that it requires excessive amounts of reserved nutrients and minerals stored in the body to do so.  They overwork the body, causing deterioration.  This directly leads to a poor quality of life, aging, sickness and death.

When you are surrounded by artificial and all you care about is social media, your music, your fast food, your computer and your cell phone, you have synthetic sickness.  You are exactly like a baby in one of those intravenous-fed bubbles from the movie *The Matrix*.  There is no such thing as coincidence.  You are rapidly losing your vital life force to a synthetic god.  You can do something about it because that is no way to live a real human life.

It is wise to protect your children from artificial things as long as possible.  In the first few years of life, they should avoid plastic and metal, which are too energetically abrasive for their little bodies.  Toys and the things they play with can be made out of wood, natural fibers and stone.[11]

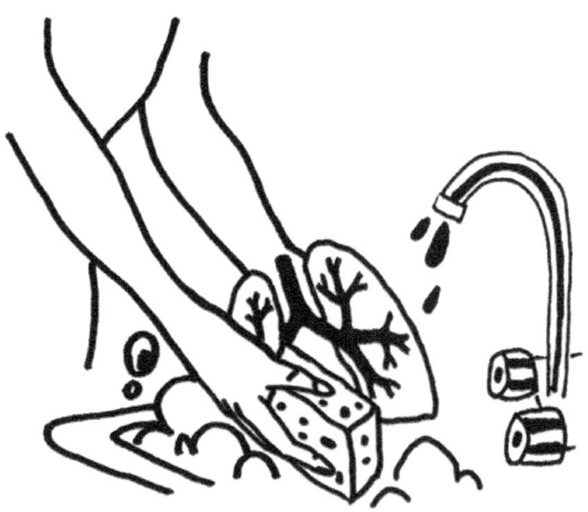

## Detoxing

Count how many chemicals you consume every day.  Is it over a hundred?  They add up.  From your toothpaste, mouthwash, makeup, hair spray, hair gel, perfume, laundry soap, face cream, hand cream, additives in the food, painkillers, pollutants in the air, exhaust from your car, gum, whiff of paint fumes, air fresheners, polyester clothes gassing off constantly, wearing band aids, working in an office with no fresh air, herbicide and pesticides in the food, new products gassing off, exposure to bus fumes, to a million other things you can come across in your day.  Having a breath mint every day is a chemical addiction.

Your organs, such as the lungs, liver, kidneys, brain, and colon are highly sensitive.  They tend to hold on to and accumulate toxins as they age.  This accumulation of nasties depletes your vital health.  You are responsible for your vital health.  Your organs cannot stay healthy unless you care for them.  The great news is that even though your liver, for example, may be hurting, with diligence it can be become better.

Do not be a fool and think that detoxing is not for you.  Your world is polluted; that is a fact.  That pollution is in your organs, causing them to be stressed and taxed.  They could use some assistance from you.

It is vital to communicate with your organs, just as you would with people.  You can talk to them.  Ask them how they are.  They will respond.  A great way to energize your organs is to smile at them.  Take a smile and swallow it.  Follow it as it travels through your digestive system, smiling at each organ as you pass it by: mouth, esophagus, stomach, pancreas, liver, gall bladder, small intestine, large intestine, colon, rectum, and anus.  This really works.

Organs hold programs from the past, from your parents, and your unconscious beliefs.  You tend to lock emotions into your organs, repressing them where they can fester.  This causes upset and damage, if you do not deal with these intense feelings.  Organs can act unconsciously and robotically if left alone.

Drinking enough water every day is one of the most important things you can do to maintain vitality in your organs.  They need enough water to function properly.  Dehydration causes your organs to not be able to do their job.  This amplifies any toxicity they might be holding on to because it makes them unable to release the toxins.  This causes the toxicity to amplify.

A great way to communicate with your organs is by detoxing them.  Toxins leave the body through your sweat, your pee, and your stool.  The idea is to clean yourself starting at the bottom and then work your way up.  In this way, you are making sure you are giving the toxins a clear run out of the body.  If you do not do it in this order, you run the risk of flooding your body with toxins.  This can make you incredibly sick.

You start by detoxing the colon.  Having an enema is a vital way to let shit go (literally).  Your colon holds on to fecal matter.  It builds up on its walls, creating a great place for festering bacteria, the birthplace of disease. You clean your outsides, why not your insides?  For your first enema use plain warm water.  For the second try warm coffee.  It is able to reach the liver from the colon, for an added detoxing bonus.

For only a few dollars, you can change your mind, let it go and feel euphoric in just a few hours.  It is not as scary as you think it will be.  It is important to take probiotics for six weeks afterwards to promote their growth and replace your natural flora, as you have just swept them out of yourself.

Then move onto detoxing the kidneys.  Bad diet and chemicals cause potassium salts to form in the urinary track.  This can slow down your peeing and eventually cause kidney stones.  They can be very painful.  A kidney cleanse is a great way to avoid this kind of trauma.

Your local vitamin shop has great herbal detoxes and cleanses waiting for you.  They do all the hard work for you by researching the beneficial herbs.  All you have to do is commit to it.  They are herbs that you normally take first thing in the morning and last thing at night.  Just follow the directions on the box.  You can also ingest a kidney cleansing tincture made with healing herbs.

Then it is on to detoxing the liver, the real cleansing workhorse of the body.  You can do a liver cleanse, also available at your vitamin shop. It is wise to ingest a liver support tincture for up to six months.  Then when you feel you are ready, move on to a liver flush.  It is an intensive flush that squeezes bile stones and toxins out of your liver.  It cannot be overstated how amazing you will feel after all of this.  You can find the instruction for a live flush online.

Finally, if you are brave, you can try hydrogen peroxide therapy. Since oxygen is the number one requirement, this therapy is amazing for your health.  It floods the body with oxygen and starts a chain reaction of oxygen production.  It can be extremely dangerous though, so make sure to do the research before beginning.  It is so powerful, it is used with cancer patients.

You use food-grade hydrogen peroxide (35%) and take it in distilled water first thing in the morning, on an empty stomach.  Start with one drop a day and work your way up until you reach a comfortable number of drops. Twenty-two is usually the max.  Do not overdo it or you will get terribly sick from toxins flooding your system.  Wait 20 minutes before eating anything.

You can do many different types of fasts, including juice fasts, or a fast consisting of lemon, maple syrup and cayenne. A raw food diet is also a great way to detox. There is so much information on the Internet concerning detoxing. There is no excuse for being ignorant.

Detoxing takes dedication. It is not necessarily easy because it removes many layers that have built up over the years. These layers are where your emotions meet your body and so it can be an intense process of revealing as you let things go. You may not release how much you are holding on to until it is time to peel it away. Watch it go!

You can be quite cranky as you are releasing unwanted layers from the past. You have to feel these things in order to let them go. It is wise to not attach to them. Be gentle with thyself when detoxing. Do not be surprised when people comment on how young and vibrant you are looking. Detoxing is so worth it because afterwards you feel and look like a different person.

## Hints for the wise

♦ Drink lots of water.

♦ Exercise daily. Preferably, not on busy streets, as breathing car pollution is poisonous.

♦ If you have body odour, odds are you are not consuming the best thing for you.

♦ Start your day with greens, a nutrient supplement and fiber like flax powder.

♦ Storing your toothbrush in hydrogen peroxide is a great way to cut out unhealthy bacteria, the main cause of tooth decay.

♦ It is wise to empower the food you are going to eat with the power of your gratitude, visualization and blessings. "Thank you for dying so I may live" works wonders.

♦ Avoid plastic water bottles as they kill the vital life force of the water. The plastic used can take a million years to biodegrade.

♦ Protect yourself and your family from toxins like cigarette smoke, paint, artificial fragrance and industrial cleaners.

♦ Avoid metal, especially aluminum, while cooking.[12] Use glass instead. (Cast iron pans, however, are beneficial.)

♦ Avoid using plastic in the kitchen. Put things in jars. Use wooden spatulas and utensils.

♦ Soy products (non-organic as well as organic) are one of the most highly processed foods and should be avoided.

♦ Plastic releases toxic chemicals when heated.

♦ Sunscreen is more poisonous than the Sun!

♦ Avoid taking calcium supplements and things with enriched calcium in them because they calcify the pineal gland.

♦ It is wise to filter your water at home and avoid drinking tap water when out in public.

♦ It is wise to eat high protein foods such as nuts, fish and spirulina.

♦ Make your own salad dressing at 2:1 oil-to-vinegar ratio. Then add anything you want for flavour.

♦ Nail polish interferes with your sensitivity.

♦ Compost your food scraps.

- Avoid canola oil as it comes from canola, which is a system engineered plant that is not healthy for you. Coconut oil is a great alternative because it is super healthy for you in so many ways.

- If you are stressed out, have anxiety or have insomnia, it is wise to quit caffeine.

- Using moisturizers and creams can be taxing on the liver. Healthy skin comes from the inside out, not the outside in. Omega oils like fish oil is the best way to have glowing skin.

- Use steam to clean your face. Boil water on the stove, place a towel over your head and put your face in the steam. Be careful not to burn it.

- Please squeeze your blackheads.

- There are stores that will refill your personal product bottles usually for a lower price. This is a great way to reduce the making of plastic garbage.

- It is wise to avoid touching toilet seats, toilet stall door handles, bathroom door handles, and faucets when in a public bathroom. You can use a papertowel when doing so.

- Moderation can work. Excessive behaviour will eventually lead to disease.

- Wash your hands first thing in the shower, before bed, before eating, before sex, after cleaning, and when you get home.

- Absolutely avoid microwaves. When you use a microwave, it scrambles the molecules of food. This kills its vital life force, rendering the food dead and useless for the body.

- It is wise to not pick berries on the side of the road, as they are laden with petrochemicals from the car exhaust.

- The microwaves from your cell phone to your head cause cancer.

- Hug trees, people and pets.  Touch is a vital part of life.

- It is wise to not to pick your nose in public.

- Go for a walk every day.

- Vitamins are processed and have lost much of their vital life force. Dead vitamins are better than no vitamins because at least you have the placebo effect on your side.  It is wise to regularly rotate your brands.[13]  Use unprocessed nutrients or eat actual fruit and veggies as much as you can.

- It is wise to avoid using crutches like band-aids, arch supports and back supports, because they enable the problem; they do not fix it. If you have to, use sparingly.

- High heels make you weak.

- It is wise to avoid consuming metal, such as colloidal silver.

- It may look like food, but it is not.

- It is wise to listen to your body.  You just have to learn to hear it.

- It is wise to get a chlorine filter for your shower.

- Sweating is beneficial for you, as it releases toxins from the body.  It is wise to do it every day.

- It is wise to take time to stand up and feel your posture.  Correct it on a regular basis.

- Avoid margarine.  It is a chemical feast that depletes your vital health.

- It is wise to not force the body.

♦   Having a healthy flow means you have a bowel movement sometime after every meal.

♦   Wearing nylons or stockings suppresses your energy.

♦   It is wise to leave food in the pod in which it is created, when you cook it.  It is a little hermetic chamber that keeps all the nutrients in.  By cutting it open with metal, you are releasing all the rich magnetism before you can consume it.[14]

♦   Organic sweet almond oil makes for a great facial moisturizer and makeup remover.

♦   Use environmental friendly drycleaners, as they are concerned about their output, and use biodegradable soaps and chemicals.

♦   Do you notice yourself getting itchy after eating wheat?  Probably means you are sensitive to it.  It is wise to avoid it.

♦   Humans are not meant to sit, especially for extended periods.  It affects your spine and your digestive system in a negative way.

♦   Your body knows what is coming.

♦   Bathe daily.

♦   Meat has rainbows in it for a reason.

Six represents the vibratory reality of alchemical consumption.

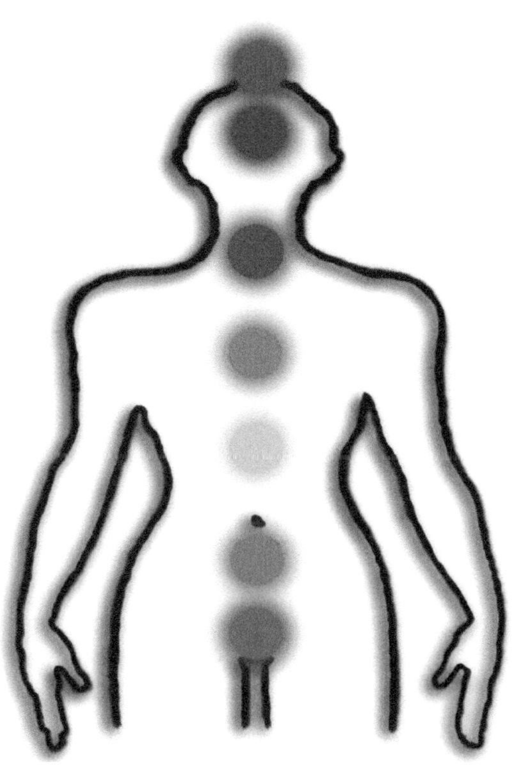



Seven is the rainbow of your consciousness apparent in personal specific
energy centres.  There are seven colours in a rainbow, humans have seven
chakras, most animals have seven cervical vertebrae, it is the number
of spots on a seven-spotted ladybug, and there are seven circles in the
seed of life.

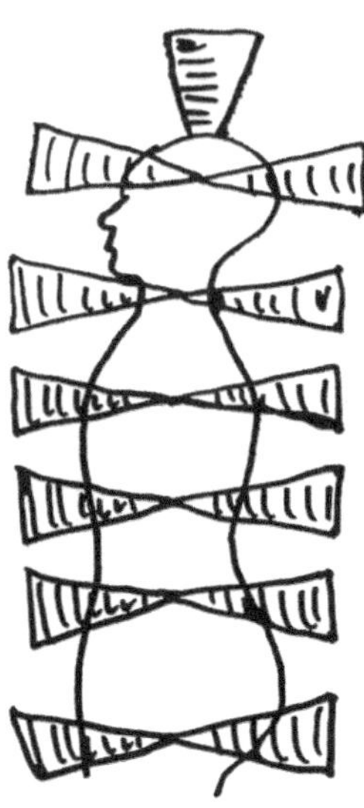

# chakra system

The human chakra system is an ancient approach to the human energetic system. Chakras are spinning coloured energy centres of vital life force. They line up along the body. Each chakra is like a flower, opening and round. They are whirlwinds of energy that express themselves at the front as well as all the way through to the back of the body. Each chakra has a front and a back.

Chakras receive and transmit energy to and from the body. They are fueled with prana from above and Earth energy from below. Each chakra has a certain energy assigned to it. At any given time, you are acting from a chakra. Motivated by money (first chakra)? Cannot stop thinking about your boyfriend (heart chakra)? Inspired by a something you saw (third eye chakra)?

Chakras are a personal experience. With dedication, intention and practice you can know your chakras. It is wise to befriend them. You can find out what they are doing. They are gateways to understanding yourself. You can contemplate the mystery of your chakras. They will lovingly share themselves with you. Your energetic health, the health of your body, the lubrication of your soul, your consciousness and your reality, for example, all have their foundation in the chakra system.

The chakra system is your energetic highway. Nadis are the channels that carry the energy from the chakras to the body. Prana flows through these channels to its specific destination. These channels can become blocked. This is the root of all disease.

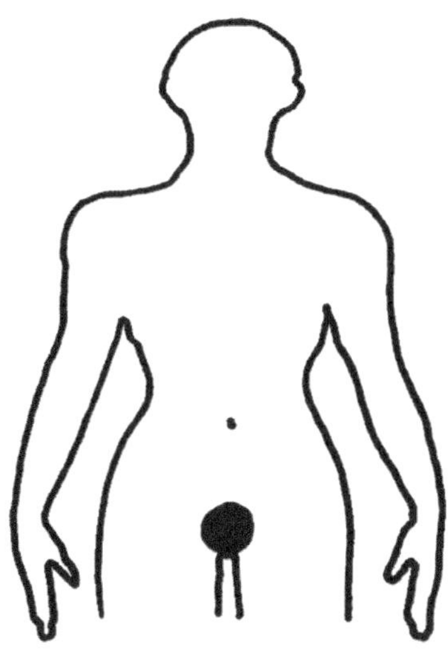

# root chakra

Location: Perineum
Colour: Red
Energy: Survival, connection to magnetism and Goddess in the centre of the Earth
Gland: Adrenal gland
Body parts: Spinal column and your kidneys
Weak: Tired and fearful
Empower: Learn a martial art
Overenergized: Aggressive and impulsive
Balance: Take a pottery-making course
Tone: "U" as in "brew"
Mudra: Thumb and first finger
Verb: "I Am"
Word: Passion
Crystal: Ruby, red jasper or garnet

The first chakra is concerned with your personal survival aspects including money, stress, fear and consumption. It is concerned with body functions including eating, defecation, the fight-or-flight response and protection of self. It is your base, your foundation and your connection to the Earth. It is your ability to be grounded and fully present in your body. The first chakra is collectively blocked upon planet Earth at this time. This means the stuck energy can be released easily if we work together to do so.

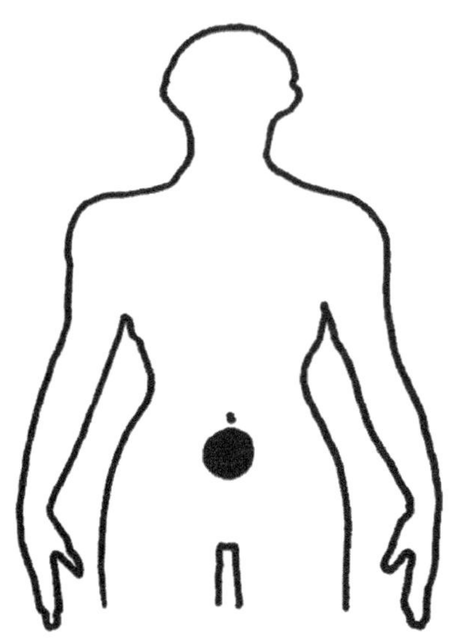

## 2nd chakra

Location: Navel
Colour: Orange
Energy: Intuition and sexuality
Gland: Ovaries or testes
Body parts: Reproductive organs
Weak: Unsexual and antisocial
Empower: Take a writing course
Overenergized: Lustful and arrogant
Balance: Take a swimming class
Tone: "O" as in "know"
Mudra: Thumb and second finger
Verb: "I Feel"
Word: Truth
Crystal: Orange calcite, carnelian or amber

Your second chakra is your instincts. It is your gut feelings and your intuition. It is your sexuality and sensuality. It is concerned with your personal creativity and ability to express your physicalness by yourself or with others. It is interested in making you feel comfortable. Your primal feelings are the language of the second chakra. It is a place of sensitivity and subtle awareness, concerned with your flow of freedom and expression. A place to hear the ancestors.

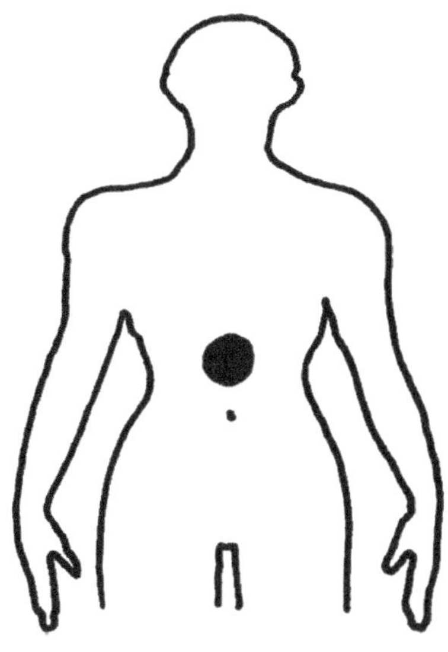

# 3rd chakra

Location: Solar plexus
Colour: Yellow
Energy: Personal power and Will
Gland: Pancreas
Body Parts: Stomach and liver
Weak: Indecisive and unsure
Empower: Take a rebirthing course
Overenergized: Bully and unrestful
Balance: Take a cooking class
Tone: "A" as in "la"
Mudra: Thumb and third finger
Verb: "I Will"
Word: Beauty
Crystal: Citrine, tiger's eye or topaz

Your third chakra is your Will. It is your ability to manifest your desires. It is your self-confidence and ability to get what you want, through action. It is concerned with defining and honouring boundaries. It is the expression of your personal passion and power. Your Will is your divine carriage. It will always guide you true. It is wise to let nothing intervene between you and your Will. Its light reveals your sacred path and the best way for you to travel.

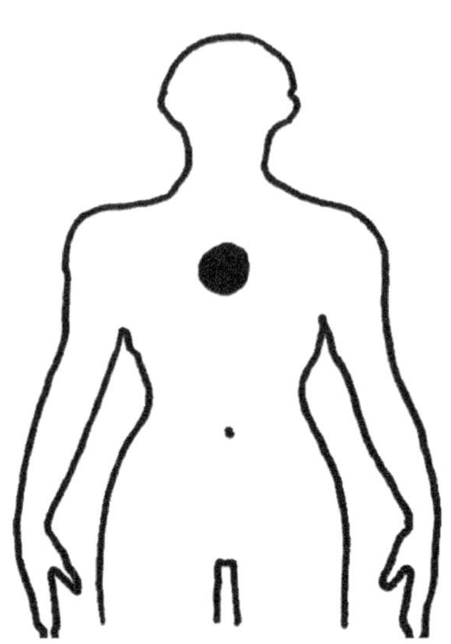

## 4th chakra

Location: Heart
Colour: Green
Energy: Love free of conditions and feelings
Gland: Thymus
Body parts: Heart and blood
Weak: Sorry for yourself and unloved
Empower:  Join a group
Overenergized: Overconfident and conceited
Balance: Take a gardening course
Tone: "A" as in "play"
Mudra: Thumb and fourth finger
Verb: "I Can"
Word: Trust
Crystal: Emerald, aventurine or malachite

Your fourth chakra is concerned with relationships and emotions.  It is your ability to have connections with others, exchanging your feelings.  It is about being open and vulnerable, sharing trust and compassion.  It is the seat of the heart.  True emotions are expressions of your personal divinity. The flow you have from your heart to those in your life is an indicator of the relationship you have with the divine.  It is wise to allow the currents of your emotions to run free and clear.  In this way, you can trust that they are pure.

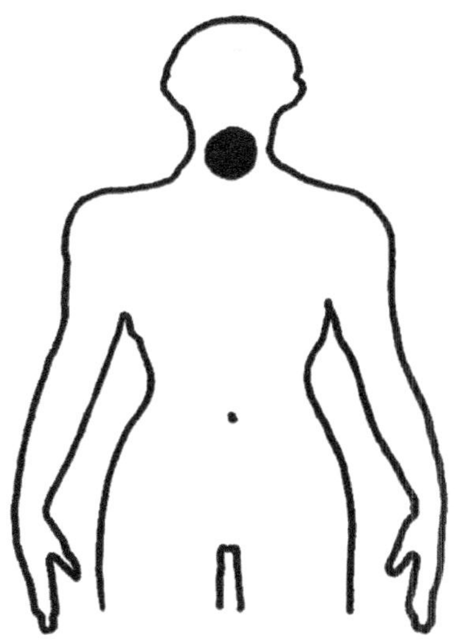

# 5th chakra

Location: Throat
Colour: Blue
Energy: Communication and integrity
Gland: Thyroid gland
Body parts: Bronchial tubes and lungs
Weak: Bullied and unable to vocalize how you feel
Empower: Take singing lessons
Overenergized: Critical and domineering
Balance: Take an acting course
Tone: "Uh" as in "duh"
Mudra: Thumb, first and second fingers together
Verb: "I Say"
Word: Harmony
Crystal: Lapis lazuli, azurite or turquoise

The fifth chakra is concerned with communication. It is your ability to express yourself with expressions from within and without. It connects your heart to your pineal gland, the gateway to the divine. It is a bridge between your inner and outer worlds. Through tones, you are able to express your higher reasoning and functioning. Your personal song is an expression of your soul. It is uniquely yours and you should be proud of it. You can do this by not holding back and allowing your expression to remain untampered. The throat chakra also has a back door.

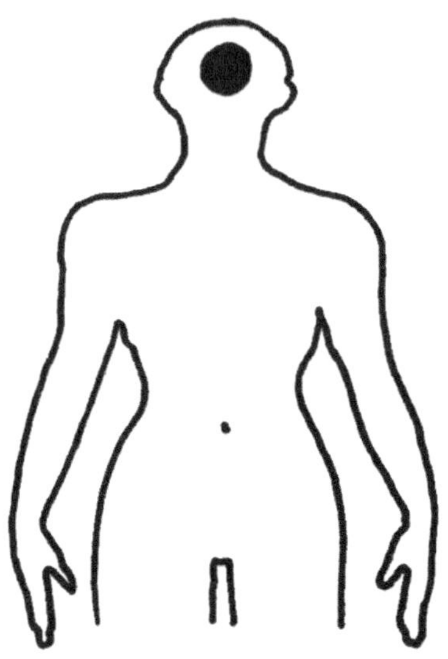

# 6th chakra

Location: Third eye
Colour: Violet
Energy: Farsightedness and inner vision
Gland: Pituitary gland
Body parts: Eyes, ears and nose
Weak: Forgetful and doubtful
Empower: Take a course to learn how to mediate
Overenergized: Oversensitive and confused
Balance: Take a weight-training course
Tone: "E" as in "bee"
Mudra: Thumb, first and second fingers together
Verb: "I See"
Word: Peace
Crystal: Amethyst, charoite or lepidolite

The sixth chakra is concerned with your visions, both inner and outer. It is the expression of your ability to take in, synthesize and express your interpretations and perceptions of reality. It is interested in deciphering input that it receives through your senses and accumulating these findings internally. It is engaged with your dreams, awareness and revelations. Your third-eye is a powerful tool. It is wise to take the time to work with it and explore its potential. You can give it freedom to share with you its ability to be powerful and insightful.

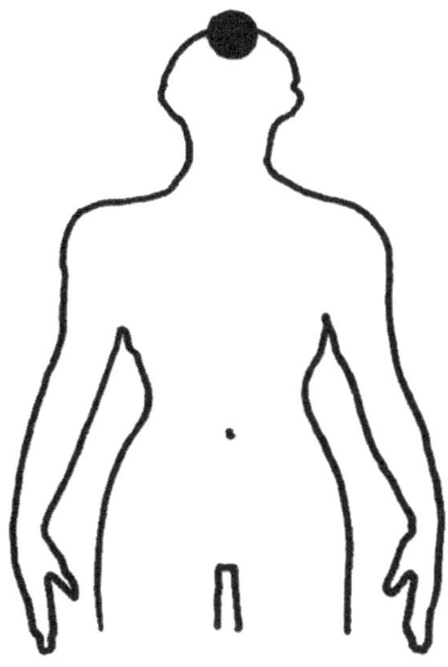

# crown chakra

Location: Crown
Colour: Pink
Energy: Knowledge, your connection to spirit and God in the centre of the Universe
Gland: Pineal gland
Body Parts: Brain
Weak: Uninspired and disconnected
Empower: Take a course on the basics of some form of spirituality
Overenergized: Impractical and spaced-out
Balance: Take a wood working course
Tone: "Om" as in "dome"
Mudra: Thumb, first and second fingers together
Verb: "I Know"
Word: Reverence
Crystal: Rose quartz, pink tourmaline or rhodochrosite

The crown chakra is concerned with your connection to the divine. It is the seat of your soul in the pineal gland. The home to your higher thought processes. It is interested in computing and synthesizing information. Memory is the fabric that holds humans together, the crown holds the jewels of your memories. It steers your ship. It is your ability to know, to learn and to think. Your crown is just that, a crown. It takes a certain amount of gusto to have the ability to carry the stars of the divine upon your brow.

## the rainbow

The rainbow is a magickal thing.  It appears out of nowhere.  It disappears just as quickly.  You are a rainbow.  You are a magickal thing. The rainbow represents the power of the fully activated, immortal human being.  Your colours come from the inside, out.  Not the outside, in.

Colour is a symbolic language with which the Creatrix communicates. Humans may take for granted how subtle of a dialect colour is, but it is a very powerful expression.  Everything you do is shaded by colour.  Colour is prose that connects to you without words.  You inherently know not to touch that brightly coloured snake.  You are drawn into a beautiful coloured flower.  You have profound feelings at the fiery sunset.

Seven represents the rainbow of energetic vortexes in the human body.

Eight is the Sabbats of the Wheel of the Year.  The ancient calendar based upon the solstices, equinoxes and their cross-quarter days.  Spiders have eight legs, an octopus has eight arms, sea anemones have eight mesenteries, the eight-spotted forester moth has eight spots, there are eight faces of the octahedron platonic solid, and animals have eight cervical nerves on each side.

# The Wheel of the Year

The Wheel of the Year is the agricultural calendar of old. It is comprised of specific holy days when Earth energy is readily available for humans to enjoy, refuel with and dance in. At these specific times, the Earth opens herself. She wants to share her energy with you.

The Wheel of the Year is the divine love story between Goddess as the Darkness and God as the Light. They dance, flow, and change throughout the year. Morphing from wee to old, they are reborn every turn. You can get to know them through their seasonal expressions. You can dance with the Goddess as a young maiden at Imbolc. Or mourn with her at the loss of her consort the Sun, as he lay dying at the end of autumn.

These are sacred days that can be observed by not working, celebrating with family and friends, feasting, a potluck, exchanging presents or having some wicked fun. Sabbats are traditionally celebrated on the closest full, or new moon to the actual date.

These Sabbats or sacred days are meant for reverence for the cosmic powers. It would be on the eve, or the night before the day, that marks the Sabbat that you celebrate. You then take the next day off to relax. Technically, a Sabbat occurs from midnight to midnight.

A great way to celebrate the Wheel of the Year is to adorn the kitchen table with seasonal expressions for the whole family to enjoy. You can decorate your home and altars with colours, herbs, images and other seasonal talismans that correspond with the time of year.

Celebrating the Wheel of the Year connects you to your ancestors. In this way you can celebrate and show reverence for the spiral of life as you honour your connection to the Earth, your sacred past, your vital future and the people in your life.

Celebrating the Wheel of the Year means that you spread the bliss evenly throughout the year, instead of cluster-fucking it all away once or twice a year. The eight Sabbats are seven or eight weeks apart, bringing joy and a reason to celebrate often. The natural support of the Universe can bring a heady delight to your celebrations.

# Winter Solstice ◆ rebirth

December 21st or 22nd. Winter Solstice can be known as Yule and Midwinter. It is the longest night of the year, the first day of winter and the beginning of the new year. It marks the return of the Sun's warmth and light. The dark gives way to the light.

It is a time to honour the darkness and stillness of the night with quiet meditation and prayer. Know that out of deep shadow is born the light. It reminds you to remember grace in times of difficulty. This is the great time of quiet, before the energetic outburst of spring.

It is an occasion to focus upon the things you would like to have born or renewed in your life. Just as the Sun calls to the seeds nestled in the dark Earth, it is a time of awakening potential in your own life. You can silently contemplate what you want to personally develop in the upcoming year.

- Traditionally evergreen boughs are brought inside to rejoice in the vitality and life force of Nature.

- Celebrate family and friends with a communal bonfire or Yule log to welcome the return of the Sun.

- Make some spiced cider you can share with your loved ones.

- ♦ Some correspondences for the season are green, red, gold, silver, holly, stars, pinecones, cedar, nuts, Yule logs, wreaths, frankincense and red candles.

# Imbolc ♦ awakening

February 3rd or 4th. Imbolc can be known as the Festival of Hearth & Home, and Candlemas. It marks the first festival of spring. The first stirrings in the "imbolc" or the belly of the Earth can be felt. She responds to the call of the returning light.

Traditionally, it was marked by the first milk of the ewes, a short time before the lambing season. This is the season of promise and potential of new life. It marks the receding of winter and the awakening of spring.

You can ready yourself for the dynamic surge of spring by cleaning and purifying your home, body and spirit. Clarify and anticipate for the promise of new growth. The world around you is active with preparations such as nest building, tree buds appearing, and baby animals being born. Flowers poke their heads through the snow.

Imbolc is a time for initiation and committing yourself to the Great Work.[15] Knowing your intentions as seeds have begun to swell and grow. You now direct and channel this energy into the appropriate areas of your life. Make plans for the upcoming year.

Focus upon the things you wish to develop and grow into. Celebrate the joy, excitement and innocence of youth. The virtue and purity of this time of year shows you that the upcoming year is what you make it.

♦ Bless your seeds for the upcoming planting season.

♦ It is a great time to pick some spring flowers and have them on the kitchen table.

♦ Traditionally a time to make Brighid's Crosses out of wheat stalks for protection and prosperity.

♦ Some correspondences for the season are white, pale yellow, pink, baby blue, snowdrops, milk, seeds, baby animals, pussywillows, cheese, a broom, chamomile and yellow candles.

# Vernal Equinox ♦ renew

March 19th, 20th or 21st. Vernal Equinox can be known as Ostara and Lady Day. It marks the first day of spring and the renewal of vitality of the Earth. It is a celebration of the awakening Earth as she is summoning and expressing fertility. Night and day are equal. Enjoy the balance and harmony of life.

It is a time to exalt the growing seeds of new beginnings you have been guarding. You begin to express these burgeoning ideas. You beckon abundance and fruitfulness to aid in their growth. You nurture them as they flourish. Buds swell and the Earth turns out many flowers. The world turns green, echoing your joy at the warmth in the air.

Night and day are in balance. This teaches you to strive for balance in your own life. Then energies of Goddess and God are equal. This is how all things begin. You make an effort to right any imbalances, by embracing both the light and the dark aspects of your nature. Call to the qualities you wish to include in your personality, to bring about this balance.

- ♦ Play checkers or chess to celebrate the balance of dark and light aspects of life.

- ♦ A treasure hunt is some traditional fun.

- ♦ It is a sweet night for a romantic date with the one you love.

- ♦ Gather a group of friends and use sage, sweetgrass, dried lavender, or incense smoke to smudge each other in order to move some energy.

- ♦ Some correspondences for this time of year are purple, golden yellow, rose, mint green, violets, rabbits, honeysuckle, leafy vegetables, eggs, jasmine, baskets, basil and mauve candles.

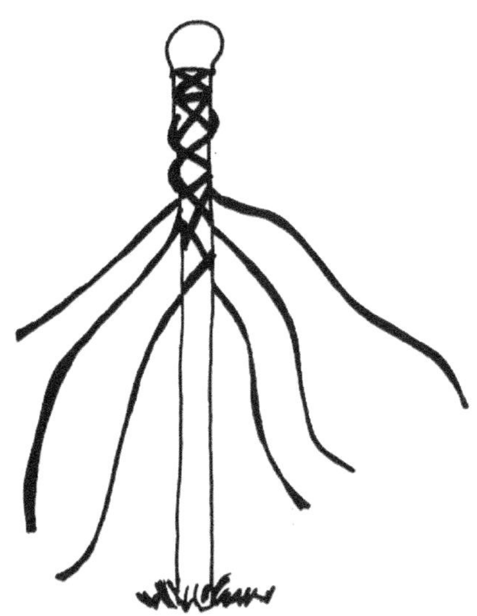

## Beltaine ♦ thrive

May 4th or 5th. Beltaine can be known as Mayday and Walpurgis. It is the final spring festival. A time when spring vitality is at its fullest expression and the Earth pulses with growth. You can celebrate the great fertility rite of life. It is a time for spontaneous joy and action. Rejoice in the blooming and procreation of the Earth's creatures.

It is an unrestrained celebration of the physical pleasures of life. You can exalt in the feeling of being alive and use this energy to fuel your creativity. Bring power to the areas in your life that you have been developing and expressing. You delight in the things you are passionate about. You can share them with others.

Through Goddess and God, you honour the mystery of love, unity and sexuality at this time. This act, known as the Great Rite, is the union between Goddess and God, or woman and man. It represents the divine explosion of bliss that is the sacred principle of all life, the coming together of two halves in order to create the whole.

♦   Traditionally people jumped over the purifying Beltaine fires.

♦   Gather wildflowers for your home.

♦   Dance the maypole with its phallic pole, and woven red and white ribbons that represent the embrace of the vulva.

♦   Some correspondences for the season are red, fuchsia, purple, lavender, cookies, lilacs, ribbons, daisies, cereals, herbal sachets, honey, chilies and pink candles.

# Summer Solstice ♦ ripen

June 20th or 21st. Summer Solstice can be known as Litha and Midsummer. It is the longest day of the year and the first day of summer.

A time to celebrate the beauty that light brings into life. Summer is at its prime and in full bloom. Mother Earth is just opening up. She is getting ready to share with you the upcoming harvest.

The summer energy is heavy and lush. It throbs with life as it flows towards completion. At the height of fertility, you can feel the change that is imminent. The light gives way to the dark. You know slowly the darkness will return. This bittersweet knowledge only lends more vitality to the rhythm of your merrymaking.

The seeds of change you have been growing are now fully blooming abilities that you have harnessed. You have integrated these ideas and concepts into your life. You now express them eloquently and wholly.

♦   Leave offerings for the Faeries on Midsummer's Eve. It is traditionally the night you have your best chance to frolick and dance with them.

♦   Take a midnight walk somewhere beautiful in Nature.

♦   Have a picnic feast at the beach and watch the sunset.

♦   Some correspondences for the season are turquoise, emerald green, blue, purple, lily, wisteria, herbs, mead, ferns, sparkles, cupcakes, garlic and violet candles.

## Lughnasadh ♦ release

August 6th or 7th. Lughnasadh can be known as the Feast of First Fruits and Lammas. It is the first harvest. Flourishing becomes fruition.

The heightened energies of the summer season are beginning to ebb. They shift from their growing time to harvest time. The shadows are growing longer as the Earth begins to let go.

Traditionally, this is a time of games and sport within the community. You share your strengths and your skills with those in your life. It is a time to honour each other's abilities and accomplishments. You gather the results of the things you have grown inside of yourself all season long. You share them with others.

As the harvesting begins, God as light wanes and becomes weaker. His spirit begins to enter into the harvest itself. Eventually he will sacrifice himself. The world will make it through the dark winter upon the bounty he has willingly offered up for the benefit of the whole. This teaches you the art and lesson of sharing yourself.

♦ This is a great time to make bread and share it with loved ones.

♦ Take a romantic walk in an orchard.

♦ Traditionally cornhusk dollies are made and burnt. They release the spirit of the corn, promoting life force.

♦ Some correspondences for the season are yellow, orange, amber, violet, periwinkle, breads, sandalwood, peaches, sunflowers, berries, antlers, dill and golden candles.

# Autumnal Equinox ✦ reap

September 22nd or 23rd. Autumnal Equinox can be known as Mabon and the Wine Harvest. It is the second harvest and one of two Sabbats where night and day are equal. Celebrate the first day of fall, with the signs of the dying all around you. The spirit of God has transformed into the food that feeds you. The harvest is climaxing and releasing its bounty.

Traditionally, this is a season of gratitude for all the Earth's life-giving bounty. Be grateful for all the blessings in your life. Celebrate with a lavish meal, and everyone dressed in their best. You enjoy the fruits of your personal harvests. You share them with the ones you love.

This is an opportune time to evaluate your life. Let regrets and old sorrows fade with the dying light. Take stock of your successes and failures over the course of the year. You know what you will release during the upcoming death cycle.

The light and dark are in balance once again. It is a perfect time to acknowledge the duality of your own nature. Notice how the dark and light play equal and important roles in your life.

♦   It is a great time to make your own wine and share it with friends.

♦   Have a potluck.

♦   It is wise to prepare for the upcoming winter by preparing your home, garden and yourself for the long cold.

♦   Some correspondences for the season are indigo, plum, brown, burgundy, grapes, apples, hazelnuts, pomegranates, roots, cider, cornucopias, paprika and dark red candles.

Samhain ♦ Death

November 6th or 7th.  Samhain can be known as the Feast of the Dead, Summer's End and All Hallows Eve.  It is the end of summer and the third and final harvest of the year.  It is a celebration of completion, regeneration and mystery.  The Earth teaches you to prepare for slumber, by drawing her energy inwards.

Celebrate the spooky nature of the Otherworld.  The boundary between the waking world and the energetic world is diaphanous.  The energies of the dead walk the Earth, visiting loved ones and taking part in their festivities.

Honour the loss and passing away of things in your life, as they make way for the new.  This time is perfect for personal reflection and the letting go of old habits or ways of being that are no longer serving you well.  Allow death to come to the stagnant parts of your life.  Contemplate those aspects of yourself you wish to enhance with the coming of the new year.

Samhain represents the end of the old year, and yet the new year does not begin until Winter Solstice in December.  This time represents the time between times.  It is not the old year, nor the new year.  It is betwixt and between.  The time out of time.

♦   Celebrate this time of year by setting out an extra plate at dinner, or a bowl of food for a dearly loved pet.  Leave a candle burning to honour the memory of the dearly departed.

♦   Build a bonfire.  Write your name on a bone and throw it into the fire (bone fire).  Divine from the cracks in the bones.

♦   Write your troubles on small pieces of paper and toss them into the fire.

♦   Some correspondences for the season are orange, black, red, grey, acorns, oak leaves, apple cider, gourds, nutmeg, cauldrons, masks, ginger and black candles.

♦♦♦

Eight represents the turning of the Wheel of the Year and living a seasonal life.

Nine is creating your world with belief.  Beliefs are as important as oxygen.  They are the lifeblood of your reality.  A human pregnancy typically lasts for nine months.

## believe

Belief is the core of life. You physically manifest your beliefs with everything you do. Your ability to believe is what enables you to directly experience life. Belief is required first to participate in experiences. Believing is seeing. You will only ever see what you believe.

The Universe will only ever give you exactly what you ask for. It is wise to open yourself up to all the possibilities of life. Accept them to experience and understand them. You can always change your mind later on.

Belief is the foundation upon which your brain builds reality. It is through this context that you are able to define your reality. For example, if you did not know what a boat was, how would you know what it was floating out in the water? Maybe you would relate it to a duck because you have no other context from which to base your understanding of what you were experiencing.[16]

Beliefs govern everything you do in life. You define your reality through your beliefs. Beliefs include your simple actions, your unconscious acts, your spirituality, your intellect, your body and your emotions. You define your reality by constantly projecting your beliefs outwards. They are reflected back to you via everything you come in contact with.

Belief is an important part of being human because it gives you fuel for life. Studying and learning propel your development. They develop your spiritual persona as you identify with different ways of being. As you grow, you find that you are drawn to certain energies or ways of believing. It is wise to explore whatever appeals to you.

There are many beliefs that you can call into your life. You can feel and delve into them to expand yourself. There are many expressions of spirit, each one a facet of the whole, none better than the other.  All equal, relevant and needed.  Beliefs are the recycling of energy throughout creation and time.  You have an ability to tap into universal beliefs.  You can learn from them and let go what does not work for you.

Beliefs define who you are.  They are the layers that make you alive.  Your beliefs are important.  It is wise to stick with them over another person or situation.  Honour them above all.  They are the core of your Universe.  Yet it is wise to be flexible with them.  Let them grow, or go if they no longer serve you.

Your beliefs run very deep.  You have been formulating them since before you were born.  They permeate your entire being.  They are so deeply rooted, you are most likely unaware of them.  They represent your unconscious, your dark unexplored depths and your shadow self.  This murky realm of your unconscious is where most of your life springs from.

Because beliefs are so important to your functioning it is wise to clean out your belief closet every now and then.  You can analyze them.  Toss them if they no longer suit your temperament.  Empower them if they are working for you.

## consciousness

Consciousness is your way of being.  Your consciousness is constantly dying and being reborn as you grow throughout life.  There are many layers of consciousness within you.  These layers are lessons.  They are layered atop one another.  As you learn your lessons, you ascend through your levels of consciousness.

Notice how people can speak the same language, but there are hundreds of different kinds of accents, even though it is the same root language. Notice how humans all over the planet are living in varying climates, different hemispheres and particular times of the night and day. All of these things are examples of how humans reside in many distinct modes of consciousness.

Consciousness is how you hold yourself. It is your beliefs, dreams, opinions, understanding, knowing, interpretation, attitude, and perceptions. It can include patterns of behaviour that you do not even realize you are doing. Consciousness is the ability of humans to be aware of their layers and layers of the energy of experience. It is the mysterious thing that weaves humans together in an unexplainable way.

Consciousness is your ability to be aware of the energy of your life and the energetic currents of your body and your interactions. It is how you live. It is the ability to see patterns in the things around you. Recognizing how you fit yourself into them. It is knowing you belong to something greater than you. It is an ability to explore yourself and getting to know your inner workings. It is your ability to express your beliefs.

Consciousness expands with knowledge. But not just knowledge of the mind, experience as well. You have the ability to raise your awareness. You can raise your consciousness through self-knowledge. Information is light. With self-awareness comes illumination to the dark shadowy realms within. This creates more layers of consciousness. This gives you natural depth. It expands your content. It brings more meaning to your expression.

Consciousness shrinks with inattention. For example, say you work 40 hours a week at a job you do not like. You are not living your life, you are working. This means that you are compromising, and only using a limited portion of your abilities. The extrasensory parts of yourself are not allowed to function because then they would cause you discomfort. So parts of you literally shut down. Your energy starts to wither and curl inward like a dying flower.

The awareness of consciousness extends into all facets of the human experience. It is something we all share. There are many layers of consciousness. There are different ways that they may be expressed, interpreted, experienced and defined. Defining consciousness is like naming all the grains of sand on the beach. There is so much detail that renders it unexplainable.

The consciousness of humans is not dualistic, as morality tries to suggest. Instead, it is a cycle, a process, an ever-growing and evolving thing.  It can barely be defined.  It never stays in the same place for more than a nanosecond.  It is so elusive as to be invisible to some.  Some humans are unable to grasp the etheric nature of their own consciousness.  It is a creature that is alive and demands to be fed.

Consciousness creates frequencies.  You have a personal frequency or vibration that is a reflection of your consciousness.  Humans all live and operate at different frequencies.  That is why some things make you feel vibrant or resonate with you, while others do not.  You know the humans on the planet are operating at different frequencies because they live in different places and ways, and they consume different alchemy.

Yet your personal consciousness is so private that it can never be relayed to another.  It is the fibers of your being.  All humans have consciousness that is totally theirs and theirs alone.  And yet all humans share consciousness on a primal level.

You can consciously raise your vibration by focusing upon empowering your consciousness.  This is accomplished through knowledge.  You will notice that you will participate less and less with the shallow frequencies of life as you ascend through the layers of your consciousness.  In fact, they do not even see you because you are vibrating at a rate they do not comprehend.

## rainbow bridge

You have a conscious mind.  You also have an unconscious mind, and a superconscious mind.  Belief helps you access these realms so you may establish a healthy working relationship with them.  Consciously making an effort to connect with these parts of yourself can explode open wonderous parts of yourself you were previously not tuned into.

The unconscious is where you store what you know, like habits, memories, your wisdom, your instincts, your cellular memory and your automatic physiological responses. Most of your actions, thought patterns and behaviour come from this deep layer.

It is wise to wake up to the fact that the depths you are consciously unaware of are an incredibly deep ocean inside of you. It is so powerful that it is running the show. You are unconsciously acting upon the programs held there. If you are not trying to direct this wild horse, you are only a passenger.

Through ancient stories, myths, symbols, archetypes and your ancestors, your unconscious communicates with your waking mind. Bringing the symbols these things carry into your reality creates a rainbow bridge of understanding between you and your unconscious mind. The unconscious does not see the difference between myth and reality.[17] It is wise to fill your life with these powerful things as a method of communication with your dark secret nooks.

It is human nature to suppress stress, fear and traumatic experiences in the unconscious, instead of releasing them. You hide things there that you do not want to deal with. Like strange ideas you formulated when you were a kid. You have many snippets of detrimental beliefs collected over your life, rotting in your unconscious like half-chewed fish. They are unformed ideas waiting to be shaped and molded by your Will into physical manifestations.

It is wise to develop a rainbow bridge with your unconscious mind. Then you have a way of communicating with it. This way you can direct your brainpower, instead of it ruling you with its unpredictable desires. These unconscious thoughts churn upwards, affecting you.

It is wise to explore your mind because it helps put these deep thoughts to rest. When you feel them fully and accept them for what they are, they will disappear.

Because you have probably outgrown these old beliefs, you can reflect upon, go through and purge old ideas, so they do not pop up at inopportune times. Take responsibility for your depths. When fear bubbles up, you then have an idea of what you are dealing with. You probably do not want immature issues from the past that have not been dealt with and released rampaging in your mind.

A great way to review your thoughts is by writing them down. You can keep a journal. Carefully go over what you have written. It is highly effective for getting to know yourself and processing your personal mire. Especially effective when upset.

It is wise to keep a working list with the qualities or lack of qualities that you feel constitute who you are. Spend the time to think through your life. Analyze the lessons you have learned. When you bring awareness to these parts of yourself, it can be enough of a catalyst to get the energy moving. If you do not like what you find, take responsibility and the time to change it.

You can have goals. If your unconscious mind does not believe you, you will not have its support. Then you will not reach your goals. You can run on autopilot and be totally unaware of your actions. But this is not a wise way to be. When you catch yourself acting without being fully present, you can follow the pattern to its source in order to become free of it.

For example, maybe you skipped breakfast, so then in the afternoon you were spaced-out. This leads to being lazy. It was laziness in the morning that made you decide to not eat breakfast. You can deal with the laziness, as it is affecting your personal evolution.

You will unconsciously recreate the same scenarios until you consciously learn the life lesson. The trick is to catch yourself in unconscious acts. Not easy in itself. With practice, you can do it. Then you are able to slow down and observe. Accept the things you do in the moment. But do not attach to them any longer. You can stop yourself and say, "I am free of feeling this way," or "I release this." When you claim dominion over these unruly unconscious programs, they will shrink and dissipate.

You can turn your friends on to this trick. Catch each other saying things that are just not right. Funny how a little bit of ridicule can work wonders for stopping old mental programs. Have your friends give you feedback on your words and actions that you were not aware of. It can work wonders for clearing the unconscious. It is wise to share with people in your life so they can assist you in this way.

For example, maybe you call yourself a loser when you make a mistake. It is not wise to verbally abuse yourself. You can catch yourself in the moment and release the pattern. You can accept yourself at times like these. Know that eventually they will pass. It is wise to let go and move forward.

Replace the crappy thought with something that makes you feel better in the moment, not worse. You do not need to focus on the things that make you feel shitty. Over-analyzing things will only make them stronger. Accept it, acknowledge it and drop it in order to improve your vital health.

You are able to form a rainbow bridge of communication with your unconscious mind. Dive into your imagination. Harness your self-power. Create a strong self-image. Sweep the cobwebs out of your cellars via active participation with your mind.

Your unconscious mind is taking in a stunning amount of information. It will only share with you what your belief allows. The rest is ignored. When you form a working relationship with it, it will share what it knows with you. This can only lend support to your instincts and your intuition. Your unconscious mind is such a powerful tool.

Developing your beliefs can be a way to process some of the suppressed emotions that reside in the unconscious, instead of letting them fester in your unexplored hollows. Belief in archetypes, for example, gives you the ability to understand your own caverns by giving you examples of how others dealt with the exact same issues in life.

Dreams are important because they are your unconscious mind communicating with you. They are real. Write them down, talk about them, analyze them and ponder them. They are your soul communicating with you. You inherently can understand their importance and apply it to your waking world.

The unconscious mind is so deep and unknowable, the only way to consciously comprehend it is through the imagination. The imagination allows the barriers that block understanding to come down. It is like a child who does not understand logic. It certainly understands stories much better. Spirit communicates through the irrational, the unconscious mind, and through the things the rational mind cannot control.

The act of creation is a great way to communicate with your unconscious mind because it gives it a voice. Anything from sewing, gardening, cooking, or sculpture to painting, drawing, silversmithing or carving can serve as a method for you to express the deep parts of yourself.

Try to catch yourself in unconscious acts like scratching yourself, touching your face, picking your nose or leaning on a body part. "Oh, I did not mean to do that." But your unconscious mind sure did. You can positively alter your unconscious programs through awareness and training.

As you can create a rainbow bridge with your unconscious mind, you can also create a rainbow bridge with your superconscious mind. Where the unconscious is hidden in your depths, your superconscious is like a vast open sky. It is your connection to the divine. It is the territory of your soul. It is the way to access spirituality, and your highest ideals.

You have an ability to know your soul through your superconscious. You can bring its power into your life by consciously choosing to act according to its guidance. All it takes is awareness. Your soul is ever present and like a bright light, wants you to follow it. You know via the superconscious what the best thing for you to do is in all situations.

Where the unconscious is like an unruly child, the superconscious is a loving grandparent. They are subtly and gently leading you down the pathway of spirit. They encourage you to take the high road.

The superconscious is the realm of higher forms of learning and expression. When you reach into the upper parts of yourself, you can discover new ways of being. When you demand of yourself more, and do not settle for base reactions, you can find your soul awaiting you there.

It can assist you in tapping into divine expressions of who you are. This can give you the tools and ability to bring more purity into your life. The superconscious wants you to raise your vibration. It guides you to ascend the layers of your consciousness, in order for you to obtain clear vision.

The superconscious offers solutions you may not have consciously known. It presents options free from ego and selfishness. It can offer up ways of interpreting reality that are based on being in service to the whole. These solutions can be opposite of what your unconscious mind gives you.

Creating a rainbow bridge between your conscious, unconscious and superconscious minds is also creating a bridge between your mind, your actions, and your soul. This kind of integration makes for a smooth ride in life. It is wise to make sure these aspects of yourself are in alignment before moving forward in life.

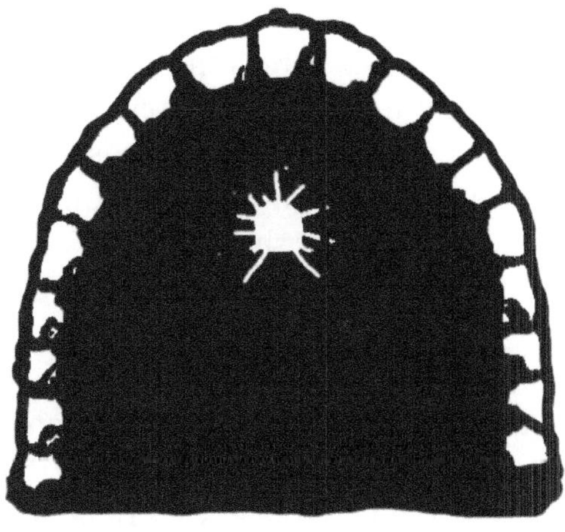

collective unconscious

The collective human unconscious is a vast, unlimited realm. It is where all of our imaginations are connected. It is where your individual unconscious overlaps with all the other unconscious' of the humans on the planet. You share so many of the same thoughts and beliefs with others. They somehow are merged together in realms not far from you. It is primal, with bottomless longings and murky desires that humans collectively share on an infernal level.

You are much affected by the collective human unconscious, even though you may not realize it. It is an energetic ocean. Your personal currents travel over its entire surface. An idea is a drop of water in that ocean of imagination. Eventually it will mingle throughout the entire population, affecting everyone.

Animals share a group soul. You see this when birds fly. As if someone gave a secret command, and they all turn at the same time. They accomplish this because their auras and their souls are connected in an energetic field or a morphogenetic field. This follows the way of Nature. A current flows through the collective. Each individual gets the same message at the same time. This is the same as the collective unconscious because we are animals too. It is an energetic field that we all share simply because we are human.

Humans, unlike animals, have individual souls. You are naturally telepathic because you are connected to all other humans through the collective unconscious. Thoughts travel easily through the darkness of primal potential. The collective unconscious holds the collective truth, memories, dreams, instincts, desires and possibilities. Everything is first created in this place. It then bubbles up, takes form and becomes manifest in waking reality.

Humans across the globe share the same stories. Time, space and language are between different groups of people and yet they still share the same stories, myths and legends. Humans carry these tales inside of them. They are shared on a core level. Humans may speak entirely different languages but still share ways of being, patterns and habits. Being human is a phenomenon that is expressed globally, not just locally.

People all over the world will answer questions with the same words. They will speak of the same feelings. They will describe the same dreams. They will recount the same experiences. They will exhibit the same patterns. They will display the same habits. They have the same memories in near-death experiences. All because humans are connected at the core level of the collective unconscious.

The collective unconscious is cosmic soup.  It is rich with hidden knowledge, secret powers and long-forgotten ways from the past.  It is full of the harmony of the days of old.  It remembers the way things used to be. It still holds all the beauty, freedom and contentment humans used to feel. It remembers us as free, dynamic and standing in our power.

The collective unconscious contains the memories of a world full of bliss.  This has always been your natural way.  This is how you know the future can be amazing.  Deep inside you still have memories of the human race expressing its unlimited potential.  They are still accessible in the collective unconscious.

The collective unconscious is brimming with power just waiting to be released.  You can connect with anyone, any time or any place.  You are intimately connected with them already in the collective unconscious.  This is why you never have to go outside of yourself to look for satisfaction. Everything is within.  The entire Universe is within you.

What you do affects the collective unconscious and everyone on the planet.  Because of this, it is wise to take responsibility for your output. Just as one car can hold up the entire highway, your actions domino into the collective unconscious.  Imagine a big tub of water.  One drop of poison has the potential to contaminate the entire lot.

If you learn new information, it creates more spiritual light that affects everyone.  This makes it easier for those who come after you to access that light themselves.  It gets easier for every successive generation to access knowledge from within the collective unconscious.  The spiritual work you do now affects the whole Universe now and throughout time.

You can access this ancient knowledge through spiritual work.  For example, when soul work occurs, it is passed from mind to mind.  You can know when it has happened in other places because you are so connected via your unconscious.  You can become sensitive to the energies within this hidden realm.  It is a primordial and ancient world.  It is not a rational or explainable thing.  It is something that must be felt or experienced in order to be understood.  These powerful and primal energies can flood your consciousness with power.  It is up to you what you do with it.

Look at the Internet as one physical example of the collective unconscious of humans.  It is a place where humans share their collective thoughts, hopes and desires.  There is no control placed upon you there.  So finally, you have a place to be free to discover your collective potential and your boundaries. The Internet is a mere tool. It is designed to show you your real power, which does not need an electric metal box to be expressed.

There is a phenomenon called the "hundredth monkey effect."[18] Scientists were studying monkeys and they noticed that when the monkeys learned new behaviour, it affected monkeys in a different locale. This suggests that once a critical number of participants know something, then the information spreads throughout the collective. In effect it goes viral. This concept is known as critical mass. When a certain number of humans know something, it spreads like wildfire throughout the human population, and it will not be long before everyone knows it.

The collective unconscious has been downplayed for so long. For most people, it has lost its ability to collectively manifest miracles. Because some people do not believe that miracles exist, they do not experience them. The collective unconscious will manifest whatever is believed. If three billion humans get together and believe something, it will happen. It is all about empowering and directing the currents of belief.

This realm is where much of what defines human nature stems from. We are barely aware of how much we as a species are motivated from this place. It has subterranean currents that flow through it, directing and affecting human behaviour. Humans unconsciously respond to its ebb and flow.

Just as the collective unconscious realms are interconnected, so too is the realm of the collective superconscious. This realm is flooded with the light of the divine. It is where we share our collective aspirations and greatest potential. It guides and directs humans via soul communications.

The collective superconscious seems to be accessible to some more than others. You could say that spiritual leaders throughout history have had greater access to the superconscious than the average human. Therefore, they have been strong spiritual leaders because they have been empowered by energy from this realm. They have been able to embody, translate and share its wisdom with the rest of humanity.

Superconsciousness represents the spiritual nature of humans. It is outside of everyday perceptions. It transcends common interpretations. It is a realm concerned with awareness and sensitivity. A spiritually centred person can access these tools within this realm in order to empower their personal evolution and therefore the spiritual evolution of all humankind. With enough dedication, you yourself can become a tool of the Creatrix and be used in the designing of the Universe.

# inner child

    The foundation of your vital health is your development when you were a child. This is reflected in the adult as your inner child. You may not want to accept the fact that your inner child is the one really running the show, but it is true. The irony of this is fun to ponder. All of your programs come from your inner child. You were taught most of the things you know at a young age. Usually three to six is when you solidified your lessons and decided on how you were going to be.

    Humans tend to be stubborn when they are wee. Your inner child reacted to their environment and made decisions that greatly affect you today. Say you were exposed to fear quite young; you then created programs to try to protect yourself from experiencing fear again. Or maybe you were abandoned, therefore you programmed yourself to always expect to be abandoned.

    The programs you created may not be wholly rational or based in fact, but your inner child does not know this. They do not care about this because they are just a child. Surely, you can remember how everything could seem so intense and overwhelming when you were little. The issues that may appear trivial now as an adult took on a great significance when you were small.

    Basically, your unconscious mind functions based upon the programs your inner child made. And in turn, your chakras carry out what the unconscious mind wants. If you are experiencing patterns, emotions and situations in your life that are causing you pain or are upsetting, odds are it is because of your inner child programs.

If you want to do something about it, you need to connect with your inner child. You need to speak with them. It is wise to open a dialogue with your inner child. You can convince them to change, alter or look at things in a different way.

You can ask your inner child to change the programs. You can have someone assist you in guided meditation to reach your inner child. You can meditate by yourself in an effort to reach your inner child. It is wise to bring awareness of your inner child to the foreground of your mind. Give them some time, energy and space in your real life. Like any small and innocent being, they appreciate your attention and will be more apt to listening to you.

Your core beliefs were programmed by you when you were a child. You can experience and become consciously aware of unconscious patterns you habitually act out, then you can find your inner child. Initiate a conversation with them and try to uncover why these programs were developed in the first place. You must get your inner child to trust you. Not so easy sometimes. It takes a calm, gentle manner.

Discuss it with your inner child. Assist them to see how the outdated programs are causing harm and doing the opposite of what they were meant to do. In this way, you can reprogram your core programs. Eventually, you will find your behaviour changes for the better as you release old patterns.

With practice, you can feel when your inner child has surfaced. Being dedicated to making an effort with them will pay off. They are like a delicate animal and it may take some coaxing them to trust you. With any friendship it takes time. But it is highly effective when there is communication happening in a balanced way. It is wise to take care when attempting to tinker with your inner workings.

Just as you have an inner child, you also have an inner mother (father), and an inner crone (sire). These are aspects of yourself that carry the wisdom of your ancestors. With dedication you have the ability to access their knowledge within. Listen quietly for their voices.

Cheesy but true, affirmations work. Just as advertising sends powerful statements that are picked up by the unconscious mind, so too can your intentions. You can write out uplifting statements and hang them where you will see and read them. In your meditations you can use a mantra that you repeat in an effort to connect with the inner child.

# the power of imagination

You learn the depth and breadth of human experience and abilities through mythos. Truth or fiction, it matters little to the soul. Beliefs empower you through the realm of your imagination. Creative thought is the language of the unconscious mind, spirit and soul. Believing in something stretches your mindscape. This ultimately expands your ability to conceive of the unlimited possibilities before you. It can make you aspire to new heights.

Your imagination is endless, as is your ability to believe. This means your access to your bliss lightning is also limitless. The only limits are in your beliefs. This is why it is wise to stretch your beliefs through learning and studying so that they do not get stagnant and start to shrink.

Your imagination is a real place. It is vitally connected to the energetic world. It weaves together your unconscious, your consciousness, your superconscious, your ancestors, your dreams, your energy, your reality, your beliefs, your faith, your soul, your sparkle power, your heart and you together with the collective human conscious, unconscious and superconscious. It is the lubrication of the soul. The soul itself is beyond your normal understanding. You need the wings of your creative mind to comprehend it.

Everything you create in your mindscape is real. It stays there, growing with your attention. This world is your diaphanous bubble that surrounds you. It coats and covers everything in your reality. It is more real and important than anything. It is the birthplace of your reality. Everything you think has its seeds in this realm. You live life through your imagination. It colours everything you do.

You can bend your imagination to your Will like a whip to a horse. Use it to propel your evolution towards enlightened living. Your creative thinking can heighten and crystallize all of your experiences in life. You can travel anywhere in your mindscape. You can give attention to anything, thereby empowering your real-time life with its fertile juice.

Your imagination is the source of your self-confidence. It catapults you through life. Bolstering yourself in your mindscape fuels your energy, which in turn fuels the vital connection to your soul. That in turn propels your personal evolution towards actualizing your bliss. This empowers your knowing of your own immortality.

The imagination is the vehicle of the soul. You can either drive a shitty, broken-down bus, or fly on the back of a unicorn. The choice is yours. Knowing your imagination is real makes the things you think about take on life. They become more powerful when you imbue them with your vital energy.

Your imagination is connected to the collective human unconscious. Therefore it has a fuel source that is unknowable by the rational you. If you allow it, it can feed your own mindscape with its unlimited potential. Creative thought develops energetic layers that coalesce into something more than your rational mind can conceive of.

The power of your Will combined with your imagination empowers your vital energy. You can become magnetic enough to manifest your desires. Pull yourself out of the chaotic rabble by defining yourself. Empower yourself via the imagination. The Universe will fully support you.

The energy of your imagination follows you wherever you go. Having highly developed creative thinking sets you apart from others. It enables you to see and comprehend things in ways they cannot. They do not have the capability to imagine anything but the most plainly obvious. Having a vivid imagination makes life fun.

Your consciousness is built with so many complicated layers. It is only through the imagination that you can begin to harness its power to understand its depths. This process is happening whether you are aware of it or not. It is wise to become conscious of your power of creative thinking, in order to empower yourself.

You can use your mindscape to create your energetic environment. Imbue your surroundings with the power of your focus. In this way, you can create fourth dimensional energetic constructs that hold energetic space.

The more you contemplate them, the more they solidify. As they grow, they offer and lend your surroundings a certain vitality that can empower you greatly.

Your imagination is a realm that is yours. You can think and daydream about anything you choose. There is no censorship there. In this realm you can be whoever you want. It is wise to go there and figure that out.

## creator of your reality

Your belief creates your reality. Whatever you place your focus upon obediently grows with your attention. Equally, it will dissipate without your attention. Everything in the Universe is energy. Whatever you feed with your energy magnifies. Your beliefs change with your intention. The energetic world will take on anything that you say and will flourish with it as long as you feed it with your belief.

Your brain takes in so much information that it has to have a way to organize the incoming info. These filters are what splice certain information out of your reality. In this way, your beliefs dictate to your brain what information is valid, and what information is to be ignored. Your brain will always try to give you only what you want based upon your beliefs. The rest of the information will fly over your head. It will be totally disregarded.

In this way, your judgments colour your interpretation of reality. You believe that something is a certain way so the Universe will give you exactly what you ask for. It will present your judgments back to you. This is why it is so important to put effort into expressing love free of conditions. If you have worked diligently at erasing your judgments, they will not keep clouding your reality with their false notions.

In essence, program the brain as you would a computer. The way you define things is the starting point of construction. Take time to analyze your personal definitions when they come up. Everyone characterizes things differently in their mind. It is something so mechanical you may take it for granted how powerful of an effect it can have on your life.

You unconsciously attract the people and situations that support your expectations. Maybe you believe you are unlovable, grouchy, smelly and not nice. It is you, yourself, who is creating yourself to be unlovable. Maybe your father was abusive. An old belief says that you as well will choose an abusive husband. So you will, unless you change that limiting belief.

The interesting thing about how you create your reality is that you become hard-wired. In a way, separate from the outside world. You are not seeing the true nature of reality, but only your version of it. For example, if you are working on the lesson of guilt in this lifetime, you will only ever see scenarios that are concerned with guilt. No matter what the people in your life do, you will only ever interpret their actions through your guilt filter. Even if those actions have nothing to do with guilt in any way.

Say your friend is excited about their exercise program and they share it with you. All you hear is how you feel like you are inadequate and guilty because you do not have an exercise program.

Your friend is working on the lesson of neediness. You had a date together but you had to cancel at the last moment because something work-related came up. Next thing you know, your friend is getting all hypersensitive and blaming you for not fulfilling their neediness. They lay on you a whole bunch of junk that has nothing to do with how you feel or what you think. The irony is that nothing you say will ever change how they feel because of their dependency filter.

When there is an accident, all the witnesses have completely different versions of the same scenario. That is because your individual beliefs colour your perception of reality based upon what you think you know. Take responsibility for your reality. You are its creator. Blaming other people because you think they made you late is dumb. You are the designer of your world and no one is to blame except yourself.

Reality is incredibly flexible. It is wise to define it how you want. With dedication, you can generate your life by harnessing your thoughts. Act like you already have the outcome you desire. Next thing you know, you will. It is wise to bring clarity to all of your thoughts, beliefs, and desires. They are the foundation of your personal manifestation.

You decide how you feel. No one can make you feel anything. No one can make you do anything. No one can do anything to you. So happens the unfolding of your life. It is completely your responsibility. You are unconsciously drawing it all to you.  Even though it may not feel that way.

You create your body via your beliefs about yourself.  This is the foundation of self-confidence.  Your mind and the power of your belief carve out your body with the way you feel about yourself.  Thoughts are so powerful they change matter.

## powers of the mind

Your mind is an incredibly powerful tool.  It is just like a super-computer in the way that it can be programmed.  Words are powerful tools you use to formulate these programs.  Programming your mind assists in accessing more than the regular percentage.

Your mind is creating your reality by its thought processes.  It is wise to have control over what you think, in order to be free of manifesting muck for yourself.  Your brain is like a hard-drive.  You program it all the time with: "I am going... I am finding… I am feeling… I want… I can…" Your brain responds clearly to these programs, diligently acting upon them.

To truly be free in the mind, bring awareness to its processes.  Trying to fight them will only empower them.  Trying to ignore them will only make them fester.  You can bring balance to your thoughts by bringing awareness to them.  When they happen, sit with them.  Feel them to their fullest extent.  Odds are, if you do not resonate with them any longer, they will fall away.

It is wise to not call yourself names when you do something wrong: "I am stupid for forgetting."  Here you are programming your brain to have low self-esteem.  That is not wise because then you will just create more of the same behaviour.

The mind can be mean. It likes to call itself names. This is not a vital way to live. You can catch yourself when this happens: "I feel this thought. I accept it and release it." If you keep your awareness on it, it will pass.

Your mind judges present situations with memories from the past. Memories of the past are subjective. They can be modified to meet your present-day requirements. They are not reliable. The world is constantly fluctuating. Experiences are different every time. Therefore, it is illogical to judge the present by the past. With practice, you can train your mind to be open. Accept reality as it is instead of packaging it in nice little parcels from the past.

The mind enjoys creating pessimistic futures. It then makes decisions based on these fantasies. "Oh, do not bother getting ready, it will not happen. Oh, I cannot put that there because it will get stolen. I won't enter the contest because I will not win." You have successfully programmed the crappy to happen. Your reality will obligingly give it to you. The future is completely unknowable. It is wise to ignore your mind when it tries to play these tricks.

All of your pessimistic thoughts, beliefs and ideas go into the unconscious. They await your ability to synthesize them, move through them, learn from them, or let them go. When you keep saying to yourself, "Oh, this is so hard," well then, it is going to be hard. All you have to do is say is, "Oh, this is so easy." Sure enough, easy it will be. Just keep telling yourself that it is painless and it will be.

Pity is a wasted emotion. Life is karma. If someone is suffering, it is meant to be. It is wise not to pity someone. Instead, appreciate that it is a difficult lesson and that they will be much stronger for learning it. Same goes for honouring someone. This placing people above or below yourself is unwise. We are all human and we all have issues and karma to be worked out. End of story.

Worry is a detrimental thought pattern. It sucks vital energy because it is an obsessive thinking pattern. It is based on things you cannot control. It is wise to be free from worrying about things you have no power over. Acceptance is the key.

"Bad" is an illusion that was created by religion so that people would feel guilty and afraid. Religion worked diligently at creating these false energetic constructs like "bad" because then many people would come and pay them money to relieve their guilt. In other words, religion profited hugely off the guilt of humans, just as psychiatrists do today. "Bad" is an illusion. Like a ghost it has no power except to scare you.

Psychodrama appears to be a major thought addiction these days. The spewing and absorbing of energy over trivial bullshit is tearing through the human collective. Chitchat and small talk seem to be a normal way of being for some. This game of fantasizing your lines, scripting your scenes and the writing of your parts, is all well and fun but draining of your vital life force.

When you talk to people at a party, to strangers, or people you just met, what are you talking about? Are you judging others, gossiping, lying, talking shit, complaining or spewing poppycock? Or are you speaking of uplifting, heart-felt things, that can assist people in their evolution? Are you talking about soul? It is wise to have control over your communications. They are powerful tools of manifestation.

Another trap of the mind is falling into victim mentality. Feeling that everything sucks, everyone is out to get you, you always lose or you are always sick. "Oh, I cannot do that because I could get hurt. I cannot say that because then he will get angry. Ah, I am always in the way. I cannot go because I do not have the strength. I am used to people taking their shit out on me." By maintaining a victim mentality, you unconsciously draw a victimizer into your reality to fulfill your victim circuit.

This victim mentality is dangerous because it does not follow the Laws of Nature, which say that you are the creator of your reality. Your thoughts, feelings and beliefs manifest physically. By believing you are the victim, you become the victim. By doing this, you are giving your power away to unseen forces. You are saying they are in control and that they are more powerful than you. This idea is utter scat.

The mind can get bored because it is overactive and has nothing to do. In this scenario, it tends to turn on itself and get hypercritical. A hypercritical mind easily falls into name-calling, thought addictions like obsessive thinking, meanness or throwing hissy fits. Your mind is alive. It needs to be fed. TV, Internet, mainstream media and movies are not enough because they have little sparkle. They can take more than they give.

Your mind needs to learn constantly to be satisfied. Regular reading, writing, puzzles, logic problems, mathematics and learning languages are healthy for the mind. The other half of your brain needs to be stimulated with music, stories, theatre, art and colour.

Your mind has another little trick it likes to play on you where you slip into scarcity thinking. Your brain can think there is not enough. It then creates reality based on this concept of scarcity. Most likely, it is leftovers from past lives and times when there was too much poverty, war and famine. You have to train your brain to realize there is enough. That there will always be enough.

The systems of the world profit off of scarcity thinking. They want you to believe there is not enough. The system uses the illusion of scarcity. But the resources are being hoarded and hidden from the general public by a small elite. A spirit person knows all they have to do is trust. Reach out their hand and what they need will be there.

The Earth loves you. She willingly provides for you. The corporations want you to believe that the resources are running out. As a human, you have a right to consume. There is no need to feel guilty for consuming, taking what you need, being alive, needing food, shelter, clothing and even gasoline. This is the relationship you have with the planet. She gives and you consume. It is natural process that follows the Laws of Nature.

Self-sabotage is another mind trick that can affect your progress in life. You have deep, dark desires that are afraid of success as much as afraid of failure. These mischievous aspects of self can pull out their gun. They can shoot you in the foot before you know it, spoiling all the hard work done. It is wise to analyze the situation when this happens. You want to learn from it so it does not happen again. It is wise to notice the pattern between procrastination, self-sabotage and guilt.

Most of the dirty mind tricks come from unconscious programs from the past. When you experience one, try to follow it to its root. Dig deep to find out why you really are that way. Usually it is a scenario when you were a kid. You can release it if you want to. Let it go and move on. Correct the thought with a better thought.

Whenever you realize you are thinking something that is destructive or not beneficial to yourself, stop the thought. Analyze it. See where it came from. What were you thinking to make you think this way? Consciously say, "I accept this thought, and release it from my mind." Purposefully think about something else. If you do this enough, you will be able to cut down on the harmful and unproductive thoughts your brain invents. Your brain is an unruly child and responds well to boundaries.

It is wise to figure out that you are totally in control of your mind. It is a tool. Like any other tool, it just needs to have parameters defined. For example, allow yourself ten minutes for any extreme emotion, as extreme emotion causes disease. All it takes is training. You can realize you can easily train yourself to achieve anything you desire.

This is the truth. Most people find it more difficult to take responsibility for being amazing. They would rather reside in mediocrity instead of doing the hard work of being awesome. If you choose to be magnificent, you will astonish yourself with the results.

## thought belts

There are millions of people who are stressed, afraid, doubtful and obsessed. So many people thinking the same way creates a thought belt, a kind of energetic conveyor belt that encircles the energetic world. When you start heavily thinking or feeling a certain way, you join the collective thought belt. There are lots of other people who feel the same way. They are thinking the exact same thoughts.

These detrimental emotions try to suck you in and drag you down with their grasping little paws. Once you are hooked into a thought belt, it is difficult to get out.

For example, say you are trying to lose weight. You start obsessively thinking about food, how you are bad, and how you have failed, blah, blah, blah. What about the other 30 million people who feel the same way? So all of a sudden you have 30 million people who all feel and think these repetitive thoughts, repeatedly. They form a cloud of wasps that buzz around in the energetic realm, feeding and growing off others like you, making everyone even more obsessed than before.

They have formed a collective thought belt. It takes on life in the energetic world. It spins around, collecting energy as it empowers the destructive thinking of all the participants. It sucks others into it. Because these thoughts are so powerful, they take on a power of their own. It is much more difficult to get off the collective thought belt if you are not aware of it.

The simple way out of a thought belt is to control your thoughts. You can recognize when what you are thinking is harmful and does not make you feel vibrant. Stop the thought and purposely think about something else. "I accept this thought, but I am free of it." With practice, it becomes easy to flow your thoughts in more blissful directions.

## ego

Your ego is your exterior identity. It connects you to the outer world. It is the bridge between your inner and outer realms. Ego in itself is not an awful thing. You use it to communicate with others. Ego is your natural defense system.

Just as a crab has a shell. Ego is illusory armour meant to be broken through so the truth underneath can be revealed. You wear your ego to protect yourself. Real friends can easily see through the protective armour. They effortlessly pierce through it to get to your meaty centre. Others are repelled by it. They will always see the shell and never be able to penetrate it to see the real you.

Your ego is your creative urge. It loves to create these little scenarios in your head. You fantasize about the ideal you. The ego is a healthy thing in these scenarios because it assists you to develop an idea or a vision of how you want to be in the future.

Once you have empowered this scenario in your mind with your ego, you can then step into your vision, in real life. You need your ego to be able to develop these parts of yourself. Your ego and your imagination together create a blueprint for your future self. They define and design your future.

You cannot have the mansion, the pool and the money, unless you can feel worthy of having these things. You have to be able to imagine yourself as a strong and powerful character, in order for it to happen. This is how the ego aids in your evolution. It allows you to imagine yourself as more than you are now. It assists in your development and the attainment of your desires.

You go through life with this hard outer shell of ego that protects you. At the same time, it can encapsulate you with your fears and self-imposed limitations unless you learn how to use it to your advantage by breaking through it.

This hard shell is rubbery. Things like wise words from elders bounce right off it. They do not penetrate into your being because your ego says it already knows and has no time for listening to the wisdom of experience of others. This attitude cuts you off from learning valuable lessons. Your overabundant ego is too inflated and thinks you are all that and a bag of chips.

This is a shame because your elders really do know what they are talking about. Usually the ego fucker does not have the experience to have reached that level of understanding yet. A fine way to be, but it takes longer for your personal evolution to blossom.

When your ego gets inflated, you can think you are something but you are not. Living life vicariously through your ego makes you look silly. Symptoms of an inflated ego are exaggeration, lying, and yelling loudly, especially yelling over someone you are talking with. It is wise to keep your ego fantasies in your mind until they become a physical reality. This way you will not come off like an egotistical dork.

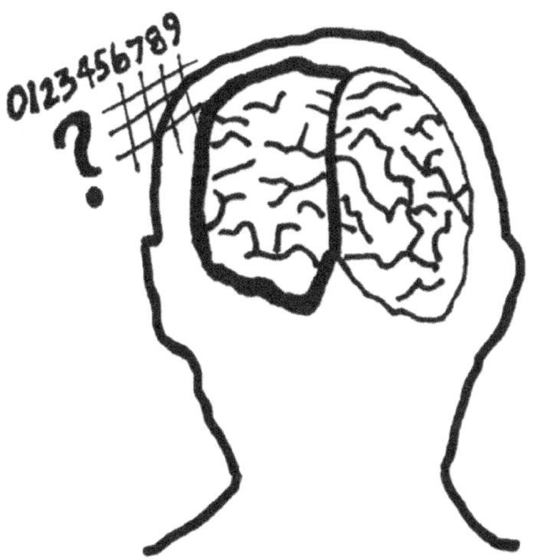


overinflated left brain

The left brain is the male hemisphere of the brain. It is concerned with facts, logic, statistics, numbers, and rational thinking. The left brain is competition-based. It fights to be right and to win. Unfortunately, it is small-minded, mean, critical and generally a bully.

If your left brain gets bloated and out of balance with the right hemisphere of the brain, you develop an unhealthy mind. The overinflated logical mind can lead to obsessive behaviour concerning greed, money and acquiring material possessions. This is how it tries to find stability. So it can reinforce its supremacy.

The left brain can cause trouble when it is puffed up because of the lack of true self-esteem and natural confidence. It tries to take over the entire mind. You can start to interpret your reality solely through the left brain. Then everything in life takes on a warped perspective. It is wise to balance your linear brain with empathy, so it does not consume you.

An overinflated left brain leaves the person thinking that they know. But the paradox is that they do not actually know at all. This leaves them totally off kilter and with a perspective that is egocentric and obtuse. It is a sickness because it is contagious and it will eventually take you down.

Your left brain works diligently to be in control. Soul threatens and scares the hell out of it. It will do everything it can to make spirit disappear when it feels its stability and its ability to dominate are endangered. It functions in a logical, linear and systematic progression whereas spirit is wild, unpredictable and chaotic. The left brain thinks spirit is just crazy. The rational mind can only understand a tiny portion of the world. It is so egotistical that it thinks it is a god.

Your left brain loves to argue and distract you from the task at hand. When it feels threatened, it will deny that soul exists. It will use every excuse to dissuade you from your spiritual path. It will fight you every inch of the way because it does not want to lose supremacy. It takes a firm Will to overcome the power of the left brain. You have to smother it systematically with love free of conditions and soulful experiences until it drowns.

Your left brain loves to use denial to convince you that there is not a problem. Denial is a sort of paradox. When you are in it, you deny it. It is a bit of a mirrored prison. Denial is the enemy of self-discovery.

Your left brain naturally locks itself and you in tiny boxes it has created with its pride. It will say, in a controlling way, "Well, I am not going to do that." Meanwhile your heart is crying, "I want to do that." These head games your left brain plays on you hurt you in the end. Usually you end up angry and lonely. The left brain has cut you off from the things you love because of its fierce pride. Cutting off your nose to spite your face hurts you the most.

To believe in the unbelievable, you have to have a brain that functions in a balanced way between the two hemispheres. You have to be logical. But you also have to be random. You have to be factual. But you also have to be flexible. You have to be rational. But you also have to be free of restraint.

The left brain has many misconceptions about life, such as a fear of dependency. Dependency upon on another is not a weakness, because it is a Law of Nature. Plants, animals, and elements all need each other to survive. Humans need other humans to survive. Everything in society is telling you to be independent. But this goes against the Laws of Nature. Separation is a major cause of disease among humans.

## Hints for the wise

There are so many areas that you can lend your belief to, so many areas that are open for study. Faeries, coincidences, the spirit world, miracles, the supernatural, ghosts, colours, elements, symbols, herbs, crystals, knots, Tantra, chaos, the moon, trees, Shamanism, meditation, hypnotism, mantras, astral travelling, Egypt, psychotropic plants, candles, lightbody, native studies, Tarot, trance, dowsing, automatic writing, Medicine Wheel, sensory deprivation, and animal totems to name a few.

♦♦♦

Nine represents the power of belief.

Ten is the illusion presented to you that is labeled "society."   The decimal system, binary system, money and time are all based on 1's and 0's.

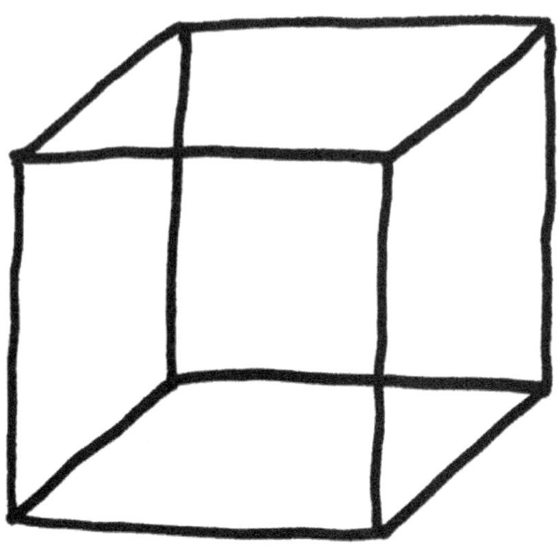

# straight lines

There is a grand illusion that is being played upon us humans. That is the illusion of straight lines. Straight lines do not exist in Nature. A straight line is truly only ever an arc of a curve. They are alien to the planet and to us. Straight lines represent the system of control placed upon humans.

These unnatural, straight grids of time, streets, money, telephone lines, buildings, schedules, the binary language and duality are fabricated illusions that have no foundation within Nature. They bind us into patterns that deprive us of our vitality. They cause stress, sickness and disease to flourish in the human race.

These things are tools that can be used properly and with integrity. But the way they are being used now creates an ugly, linear mess. Somehow, straight lines have become our master. We have given our power to them. In the process, we have lost sight of our true nature. For the real power of the human resides in the spiral.

The system in an effort to be efficient has worked diligently in labeling, organizing, categorizing, separating and standardizing everything in our world. Every square inch has been measured. A price tag put on it and an owner assigned.

The system believes it does this for our own good. At least that is what it tells us. The control certainly creates lots of profit. The trouble with these tactics and their square methods of command is that they run opposite to the spiralized, curvaceous and seemingly chaotic nature of Earth.

# the system

Society is an illusion. It is a controlled system, a fabricated creation designed to make humans weak, afraid, docile and easily malleable. It is wise to realize the system does not have our best interests at heart. Society is designed to make us live in terror, to hurt us, to make us fat, to make us stressed out and to make us slaves. This is how the system profits.

The foundation of society is a tradition based upon the subjugation and defilement of women, children, the Earth, animals, plants, resources, boys and men. It is enforced by a mechanical, heartless, and soulless system.

It does not care about the welfare of anyone, only itself. The traditions so casually accepted as inherent are fabricated byproducts of a corrupt and malicious system. It is a robot, a bully, a computer, a ghost, and nothing more.

Etiquette is the energetic deputy of the system. It tries to keep us all in line with its rigid and outdated modalities. It thrives upon the suppression of feelings, which are the foundation of being human. It is wise to ask ourselves why the system is asking us to hold back.

The system is an out-of-touch parent telling us to go 30 kilometers around the corner, when everyone always goes 50 km. Why is etiquette, with its binding ways, expected to override our natural Will in life? We are meant to comply with a set of unnatural laws that the system has fabricated to govern our behaviour. If we fail to comply, we are punished.

The entire system is one in service to greed, profit and power hoarding. It is a system in service to the individual. Not to the whole. Because it is not in service to humans, the Earth and her creatures, it does not concern itself with their well-being. Therefore, it hurts us with its apathy.

Unfortunately, the system ceaselessly works to lower the vibration of the human population. It does this in order to maintain the illusion it is projecting onto our consciousness. The system wants to fill our heads with crap. It wants to turn our heart power way down so we do not realize the grand chimera it is producing and slip out of its control.

The system teaches us not to trust ourselves. But instead to give our ability to decide over to an authority outside of ourselves. This makes us hesitant to have faith in ourselves. In essence we become dependent on the system to take care of us and make our decisions for us.

The system does not want us to know that the Creatrix is inside of us. Then we would realize we do not need the system with its middlemen including doctors, insurance brokers, lawyers, judges, police and corporations. The system thrives on our guilt, fear and stress. It does not want us to trust ourselves. But remember, it is the system that is fucked up, not humans.

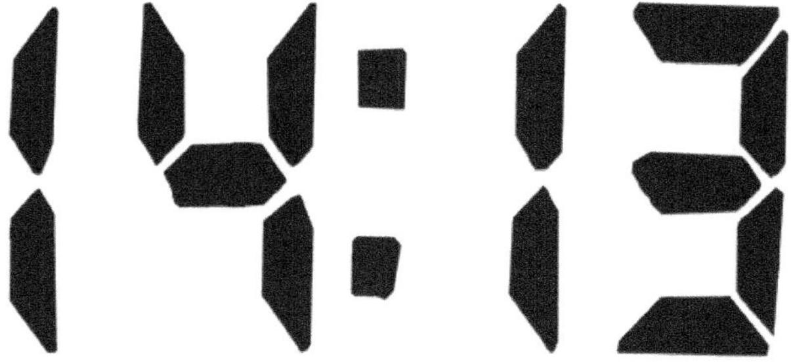

## time

Clock time is an unnatural, fabricated phenomenon that enslaves many humans. It is two-dimensional with its logical sequence of numbers, one following the other in a straight line. It is something that does not exist in Nature.

The sequential slaughtering of our night and day into little tiny units has a rigorous negative effect upon our psyches. It heightens our obsessive-compulsive natures instead of our harmonic ones. How many times are we wishing our lives away in order to make time pass to a better time? You cannot ever get it back.

Time chooses the logical, left brain. It tells the chaotic right brain to take a hike. The backbone of our whole society is automatically out of balance because of this exclusion of the irrational part of our natures. We can never reach our true potential if the foundational concepts of reality are incomplete and out of balance. Dictated decisions like clock time keep us limited and small.

Real time, as in the Sun rising and setting, the moon rising and setting and the ebb and flow of the tides is spiralized. These events flow. They are different every day. They spiral chaotically through their cycles. They are not perfect. Therefore, they are not dependable in the eyes of the system. So the system modifies them. It fabricated a grid of time that is neurotically forced down upon us.

The Earth's natural time of day and night constantly shifts depending on how fast or slow the Earth is rotating. But she rotates at different speeds depending upon the Sun's location. She spirals. Therefore, appears unpredictable and chaotic.

The system has developed vastly intelligent computers using atomic timekeeping devices to try to keep up with her. Apparently, it is a difficult task. They constantly have to make up for lost time due to the perpetual fluctuating motion of the Earth as her natural timekeeping is affected by the Sun, moon and other planets. Must be like chasing a firefly.

The clock has become the master for some. This inorganic pattern of compliance causes vast amounts of stress for the human population. We struggle to conform to its inflexibility. The constant rushing to fill the expectations of time causes the human's energetic system to be taxed. Our immune systems are weakened by the constant stress to adapt to an artificial schedule.

The way we spend our time is governed by rules such as days of the week and work schedules - usually daytime working, nighttime relaxing. This is dictating to us how we live. It is the blueprint for the human schedule.

Sleep from ten to seven, eat at seven-thirty, work from nine to five, eat at six, drink beer at seven-thirty, TV at eight, and sleep at ten; repeat. People are living for the weekend because they are stuck in a nine-to-five, five-days-a-week grind. It is a cliché because it is a sad truth.

This pattern of conformity is not how humans are supposed to experience the world. We are not meant to have every nanosecond measured for its viability. No, we are not robots or machines. We are not meant to be plugged in. The stress caused from this kind of lifestyle is killing us.

Humans give their power away to time. It has become more important than our vital health. Religiously following the concept of time creates an unbalanced lifestyle. It negatively affects our natural rhythms.

Rushing for time is a real problem as it is unhealthy for our bodies. We are ever striving for goals outside of ourselves. No wonder there is rampant stimulant use as humans try to keep up with the clock. Burning the candle at both ends shortens our lives.

Schedules are square. They are composed of linear, repetitive and sequential numbers. They put the square in our brains, when our brains are a whole bunch of spiraling circles. This unnatural linear patterning causes discomfort and weakness. It goes against our natural way of being. The human race is constantly caught off guard. In this way, our energy is way easier to manipulate. We become like deer, easy takedowns.

Daylight savings is an example of a system decision concerning time that screws with our circadian rhythms, our natural clock. It is an example of how for efficiency the system has decided it is best for the clocks to change.

But it is the worst choice to be made for humans. It may be logical for the system. But it is not logical for the human body. Daylight savings majorly disrupts our routine and causes sleep cycle turmoil. It forces us to compensate for the shifting hours. This is detrimental for the health of humans. Try explaining daylight savings to your cat when his dinner is an hour late.

The Creatrix knew what time it was when night and day were created. We are naturally atuned to her cycles. The system does not have the know-how to manipulate our relationship with the dark and light. They do not have the right.

Traditionally, humans have used a lunar calendar with its 13-moon yearly cycle. It is in harmony with the moon blood cycle of women. The solar calendar of 12 moons per cycle is contrived. It causes us great strain deep in our unconscious minds because we are trying to conform to an inorganic pattern instead of being in harmony with the natural flow.

Humans are collectively stressed out and in survival mode. As a result, when calendar holidays created by the system come around, we celebrate too much. This excess creates destructive patterns concerning addiction that last just long enough until the next holiday comes around. Tradition binds us tightly.

It is difficult to push against collective consciousness. Everyone is conditioned to participate in rituals based on the system's solar calendar. These patterns directly create detrimental generational patterns of disease in the human race.

INDU  -777.68  VOLU 2.033-8265
INDP  10365.45  UUOL 74.601.70
NX*   -686.14  DUOL 1.959-5145

## money

Money is an energy that has enslaved the world. It has taken on a life of its own. It has made humans its servants. It is square because it is a linear, progressive set of numbers. It has no foundation in the Laws of Nature. Fabricated equations have been assigned to every facet of existence. Every transaction that occurs is quantified by this inorganic measuring system.

Everything in the world has been broken down, categorized and has a monetary value assigned to it. This grid-like division of life and labour has created a specific, digital way of thinking. Everything and everyone participates in this standardized way of assigning value to everything.

Real estate, products, time, labour and services all have a fabricated, mathematical equation affixed to them. It is the source of this mathematical equation that needs to be examined. It is inherently flawed and not suitable for humans.

The ability for the system to correctly determine the value of reality is skewed by its own cubic nature. How can a straight ruler correctly measure an orange? Much of reality is falling though the gaps that are created by measuring round curvaceous things with linear rulers. This falling away is our heritage. It is our collective soul and our ability to believe, to feel and to know.

Our spirit is falling through the cracks of our reality because of these gigantic decisions that are being made by the system in an effort to govern humans. These decisions are not correctly defining what is important or not important to us. Our world is in service to profit. It is not in service to promoting the soul health of humans.

Money dictates our reality more than anything. For example, money says the stock exchange is valuable. But public schools are not valuable and therefore can be sacrificed. Money dictates that football is important, but feeding starving children is not necessary. It says movies are critical, and yet adequate housing for homeless people is negotiable.

The binary code is a language based upon 0's and 1's. Money is also based upon a mathematical system that is centred on zeros and ones. This concept of 0's and 1's is a logical system that has no foundation in Nature. It is dualistic and contrived.

This code is a cyborg code. Why would humans allow a system outside of themselves to use a cyborg code to manage them? Why would humans not use a human code to manage themselves? The cyborg code does not have the ability to effectively govern  humans and our natural world.

We take it for granted that this square mentality has taken over our perceptions of the world around us. This kind of dictated, standardized thinking weakens us as a species. It cuts off our ability to be able to think in spirals. Spirals are our inherent way. Freethinking enables humans to evolve into patterns that benefit the whole instead of serving the few. The concepts of time and money make us calculating, not accepting.

The monetary system is the foundation for all systems of the world. It has been clamped down upon society. The system wants us to believe it is the only way to function. But this concept is an illusion. Society could function in other ways other than the monetary system.[19]

As an example, humans had a solid trading system in place for millennia. Today there are places of bartering that we can join. We can barter and trade our resources and skills for other resources and skills. It is an amazing and viable solution.

The monetary system raises interesting questions. How did the land come to be divided? How was its ownership assigned? What about the white people stealing the land from the native people with false contracts and treaties that were never honoured? What of the division of resources? How do the corporations have the rights to the resources? What legal standing do the governments of the world have to charge income tax? Why do less than a handful of families own most of the world's resources?

Sparkle is now quantified in order to create profit from it. Traditionally, profit was measured in tribal quantities. There were no divisions within the tribe. Everyone shared everything. The richest person was the one who gave the most away. The whole world shared and profited equally and together.

There was so much for everybody, the concept of personal profit was never heard of.  Destruction of the tribes and the dawn of individuality changed that.  War and profit go hand in hand because of the violent need to protect one's profit.  These things are not in service to the whole.  They are in service to individual greed.  Money is cheap.  Freedom is expensive.

Money itself is energy.  It is neutral.  But when people get obsessed by it and focus upon it above and beyond everything else, it becomes a problem.  It becomes an addiction that can deprive you of your soul.

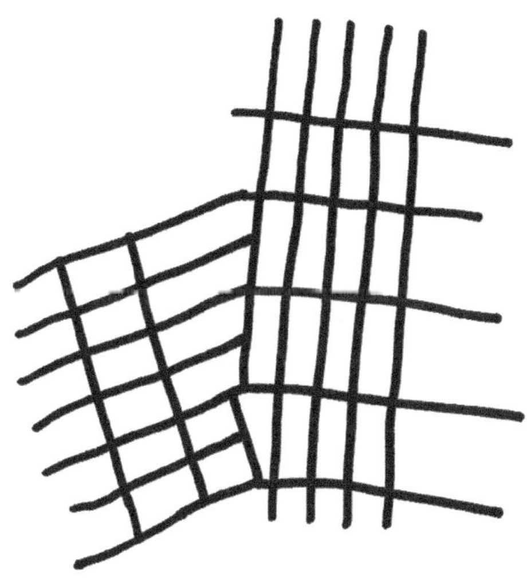

# design

The majority of architecture and city planning are based on straight lines.  Everything in the Universe grows according to the spiral of life and sacred geometry, except society's architecture and city planning of today.  Humans live in square boxes in a circular world.  That is the opposite of wise.

The stop and go of traffic lights and the straight lines of roads brainwash our brains into a set and limited way of being.  This mode of thinking is cutting us off from accessing the true and natural flow of our brains.  It is making our energetic systems degenerate.

Buildings are designed with straight lines that create corners.  Corners are unhealthy for humans because they cause our energy to be stilted.  Then we are not able to reach our maximum potential.  Energy travels in toroidal or donut form.  When it hits a corner, it goes kind of funky, and off its normal course.

This constant jolting of our energy makes our energetic system work harder to compensate.  It stresses us out, ages us and suppresses our immune systems. This is the opposite of environments that empower us and lead us to know our immortality.

We are magnetic by nature and we imprint our surroundings. When we imprint architecture that is square and totally contradictory to our true natures, it weakens our electrical circuitry. This causes us to be unable to reach our full potential as humans.

The designs themselves deprive us of vitality. They squeeze the soul out of us. They limit our ability to express ourselves. It is like putting a bird in a box. There is not enough room for him to flap his wings and therefore he cannot fly. How can we know if we can fly, if we are never given enough room to stretch our wings? Maybe that is why temples of old are always so vast.

Corners create stagnant energy, which creates dust, which creates grime, which creates bacteria, which creates sickness. In the forest, there are no corners and no separation. Germs do not prosper the same way as they do in corners. Energy is meant to flow. Corners are the opposite of flow. They trap the energy and make it fester.

This festering energy has a great negative impact on our energetic systems, our health, and our lives. This becomes obvious as we can see by the rampant disease caused by simple energetic blockages. Not having square architecture in our lives would make so much disease simply disappear.

Metal interferes with the flow of energy. Vast quantities of metal interfere with the energetic systems of humans. It interferes with our flow. So crisscrossing metal makes it difficult for humans to hold a charge. It challenges us to be grounded, become enlightened or ensoul ourselves. Too much metal affects our auras in a deeply negative way. It does not allow them to expand properly.

Cities are huge depositories of metal. They interfere with the vital health of humans because they override our natural circuitry. They cause our energetic systems to be out of alignment. This causes cells to mutate, thereby causing disease. Cities right now hurt humans because they are designed for efficiency, not human ergonomics.

Cities create dead space. Dead space has no vitality, magnetism or sparkle. It has little or no life in the form of fresh air, trees, oxygen, plants, animals, vital energy or anything natural. It consists of buildings, concrete, corners, metal and electricity that create lots of death (and not the fun kind).

We imprint our external reality by the nature of our inherent magnetism. Our magnetic field flows outwards and then back to us in a toroidal form, bringing with it the energy of our environment. When we look around and all we see is asphalt, concrete, metal and wires, this is what we are imprinting. These dead spaces are lethal for life. We need the vitality of Nature in order to survive.[20]

Our bodies go numb when we participate with these places that are energetically dead. This numbness penetrates our being, making us cold. It makes it difficult to ensoul ourselves. It leads to dullness and apathy, which makes us unable to care. This uncaring spreads and leads to killing and more uncaring about the killing. High levels of insensitivity and paralysis due to too much dead space are a reason the world is the way it is.

Sacred geometry is the foundation of creation in the Universe. It is the use of simple shapes to build complex structures. It can easily be applied to architecture and city planning. There is no reason why sacred geometry cannot be brought into our living environments and cities to make them healthier for humans. We could create architecture that was able to assist us in having healthy energetic patterns, thereby promoting vital health.

The idea is to live in buildings that are designed to resemble biological structures. Homes, for example, can use natural building materials that promote life force. In this way, they create an electrical field, something that humans need in order to survive. Our energetic systems are being severely damaged by the lack of life force in our living environments.

Buildings of old, like the temples of Egypt, were based upon harmonic proportions. They brought vital health to the people because they were alive. These kinds of buildings promote soul and bliss. The ancient Egyptians would carve these incredibly large blocks of stone so the temples did not contain even one corner. These buildings are still piezoelectric even after thousands of years. They are alive. Even today they still promote vital health in humans.[21]

Furniture designs are based on straight lines. They can cause tremendous pain in humans because they are not ergonomic. Furniture is not the right height. This causes pain and long-term damage for people because it does not fit their body type. It affects our spines and digestive systems in negative ways.

Consider how much time we spend in chairs. Who are the designers of furniture? Do they know what is right for the human form? The repercussions are far-reaching. We are damaging our genetics with poor posture because of clumsy furniture design. The design of the angles is incorrect. The concept of one-size-fits-all in furniture is bizarre and off the wall, as humans are as varied as popcorn.

## electrical grid

The fabricated electrical grid that surrounds the planet is affecting the energetic systems of humans in a negative way. The electrosmog creates a net that cuts us off from receiving higher vibrational information, by interference from synthetic waves. This throws off our natural ability to receive and interpret energy received from our souls, Goddess below and God above. The electrical grid interferes with the natural expressions of our auras. We cannot unfold ourselves properly because of electrical interference.

This electrical grid causes cancer in humans because it overlaps the nervous system, causing cells to mutate. Too much electricity fucks with our personal, natural magnetic and electrical fields. This is a cause of disease. Electromagnetic frequency (or EMF) is the electromagnetic field produced by objects charged with electricity. It has permeated our living environments. It has become completely unavoidable. Its superfluous presence is killing us.

This abundance of electrical energy is inorganic. It disrupts our natural energetic rhythms with its pervasive intensity. It uses us as a way to ground. This stresses our systems as it passes through our bodies. In human history there has never been this amount of electricity or these high rates of cancer happening. This is not a coincidence.

All the radio waves, cell phones, Wi-Fi, power lines, radio, TV, computers, refrigerators, telephone wires, compact fluorescent light bulbs, X-rays and microwaves are negatively affecting the health of humans. They weaken our immune systems. They cause us to become sicker, easier.

Humans have given so much power away that we have allowed the system to create living environments detrimental to the well-being of our race. This is out of balance. It will only get worse until humans collectively take back their power. Humans need to put their needs first over the system's needs once again.

Electricity is left uncontrolled. This creates electromagnetic fields that radiate in the environment and through the wiring, affecting everything in the vicinity. The issue is the transient pollution that electricity throws off. These transients can infiltrate everywhere. They follow the wiring all the way to the power lines. So our neighbour's electricity can negatively affect us, as well as having a cumulative effect. It is wise to avoid touching or being near electrical devices when they are hot.

A wise person sleeps with no electrical devices plugged in near their beds. It is wise to put the alarm clock as far away from the bed as possible. Unplug electrical devices when they are not being used. When something is plugged in but is off, it stills emits EMF.

EMF directly affects the magnetism of the human, causing it to be disrupted, weakened, and easily dissipated. This depresses the aura, leading to a vulnerable immune system, equaling sickness. It would be wise for humans to wake up to the fact that technology left unchecked is killing us. It also mutates the growth of our cells by overriding our nervous systems. This makes our insides squiggly. We have no way of knowing the damage that is happening to us until it will be too late.

There are ways to reduce the amount of EMF coming from electrical outlets and cords with the use of magnets. You can buy a magnetic roll from craft stores, used for making fridge magnets. Unscrew the face plate from an electrical output. Cut an 1½ inch strip off the magnetic roll and place it on the little metal box of the socket. Replace the cover. For cords cut a piece of the magnetic roll about the same length and tape it to the outside of the cord.

Funny how we are told that technology is growing so fast we cannot keep up with it. And yet the wiring above our heads that links all our houses is the same as it was back in the late 1800's. That is over 135 years of the same wiring that is so limiting and damaging to our human energetic systems. It is wise to think about that one.

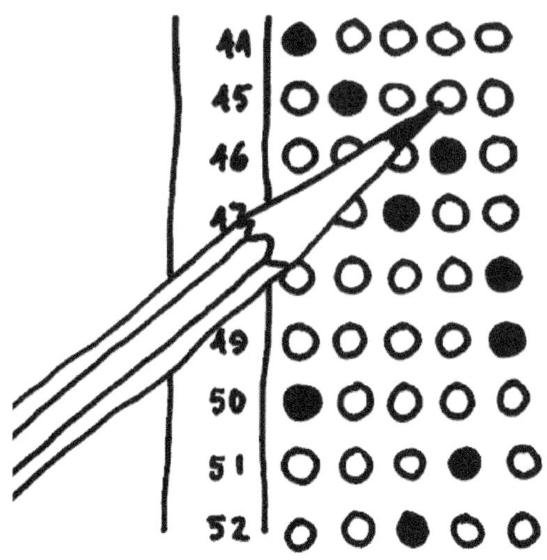

## standardization

Standardization is creating norms. It makes everything uniform or consistent so it may be coordinated for efficiency within organized bureaucracy. Like snowflakes, every single human is unique. We are all special. We all have individual needs. Standardizing reality is an attempt at regulating what it means to be human.

We all see life completely differently than our neighbours. Our various experiences, ideals, and expectations make us special. We react, learn and feel differently than others. Our bodies are singular and particular. We all think and process thoughts differently. Everyone has their own way.

By trying to make everything the same, we are becoming collectively numb. We are losing access to our collective soul. The human soul is a unique spark that cannot fit into a standardized mold. By trying to force us to comply, the system is trying to make our souls lose their vitality. We are being forced to conform to square molds that are utterly alien to our spiral species. This is causing untold stress on us. Stress leads to disease, disease to death.

Standardizations lead to clothes that do not fit, chairs that are uncomfortable, toilets that do not flush at the right time, and places where everything is white. When things are made normal and the same, what it does is make everything boring. It excludes the quirky nature of life. It excludes the ability of life to change as it flourishes and grows, learning from the past. If left unchecked, standardization will have us wearing beige and living in rooms painted mint green. Might as well be a mental institution.

Standardization is a fanatical effort by the system to control the human population. The system is unrealistic when it discounts things like soul, emotion and the individual. Neglecting our consciousness makes it fade. By ignoring these vital concepts, it is suggesting they are unimportant in our lives.

Standardization takes the focus off these critical aspects of what it means to be human. It makes them grow less, instead of more. It is wise to ask why the system wants us to have weak souls, little emotion and no individuality.

These issues are the bones of humanity. It is wise to guard them with our lives. Our uniqueness is a jewel to be treasured. Not to be discounted because it is inconvenient to organize and systemize. This blatant disregard of our individuality is a real threat to humanity. We are letting an inhuman and artificial system make subtle yet drastic decisions for us.

Standardization is taking away the biodiversity, for example, of the food we eat. There are thousands of different kinds of heritage vegetables that are not in the grocery stores because they are somehow not profitable. The food in the grocery stores is there because it has the longest shelf life, in order to increase corporate profits.

The system is telling us to fit the mold instead of being ourselves. This makes us unsure of who we are. It causes neurosis in the population. We are experiencing a collective identity crisis. This weak foundation is making the world suck right now.

Standardization has penetrated the concept of beauty within society. To be considered beautiful we have to fit into a standardized mold. If we do not then we are not even considered. What happens to all the humans that do not fit into the mold? They are usually labeled as ugly, criminals, mentally ill, street people or drug users.

These people are spirit people who are possibly too intelligent to conform to the mold of standardization. They would rather maintain their free will even though it means living life at a different level. Society runs them over like a steam truck. It makes them out to be lesser.

The system intentionally makes it difficult to function if we do not conform to the rules. The question is, who is making up those rules in the first place? Why does it have to be so detrimental to the health of humans and all the creatures of Earth?

Standardization makes rules that are absolute based. It says everything is either black or white. But we know this is not the truth. Realistically life is various shades of grey that are constantly shifting and growing. There are no absolutes.

Standardization makes everything so pale. Where are all the colours? Look at the marvelous colours Nature puts out to make us feel blissful. Nothing is more beautiful than a deep red rose. Real life is vibrant, potent and colourful. Vibrancy comes from variance. Variance is based upon the Laws of Nature and the spiral of life.

## labels and categories

It seems in society today everything must be labeled in order to be understood. A person and their unique ideas are thrown under a label even though they have nothing to do with that label. If someone speaks of ideas where humans work together and share things, they are labeled as communists. If someone speaks about a higher power, they are labeled religious. If someone speaks of female empowerment, they are labeled a feminist. If people work in a spiritual group they are labeled a cult.

Labels are a form of standardization that is dangerous. It deprives humans of the right to implement new ideas. This categorizing of ideas debases fresh notions. It throws them down the tunnel of tradition. These labels can make us feel hopeless and unable to get out from the heavy umbrellas of the same old story from the past. Unfortunately, most labels are negative. They have negative values assigned to them. Especially if they threaten the system and the grip it has on an idea.

New ideas threaten the stability of the system. It does everything it can to debunk them. As an example, look at the word "conspiracy theory" and the eraser effect it has on new concepts. The system only supports its established ways of thought and lets everything else swim underneath the dock in murky waters.

Ironically, it is these new burgeoning ideas that would have the potential to create solutions to some of today's merciless issues that are causing so much stress and damage to life. It is wise to investigate things that are labeled as controversial, flakey, unproven, hokey, New Age, anti-Semitic, feminist, or as a conspiracy theory.

Only a mad scientist thinks it is possible to label and divide everything in life. Life is constantly changing, growing, evolving and spontaneously regenerating. It is evolving much more than is being acknowledged. There is so much going on that cannot be defined, let alone quantified. There are so many layers to our ability to be expressive that cannot be captured by the left brain, as many of these expressions are based upon soul, emotions and energy.

Labeling, dividing, dissecting, and separating everything kills spirit. The spirit of life comes from its cohesive nature and its interdependency, to make something more than just the parts. Splicing out the fun makes for numb, obedient humans. It kills the urge to be spontaneous.

There are too many shades of grey for such divisions to correctly reflect the fluctuating trends. By ignoring the shades of grey, we are ignoring a vast percentage of ourselves. This can only lead to disaster. It limits our ability to believe in our own creative potential. It shortens our stride and cuts away at our growth. It limits our perspective and shrinks our desire for more.

## separation

Labeling everything creates separation. We are divided into artificially created categories of different types of humans. We are divided based upon colour, race, language, social standing, financial standing, family name, bloodline, education, where we live, our job, the car we drive, our possessions and the clothes we wear.

We are cut apart. This can make us lose sight of the truth. Which is that we are all human. Separation creates conflict. Conflict is how the system profits. Separation is a tactic of war. It divides humans and causes warfare. Humans are not the enemy of other humans.

Separation causes pain in humans because it is so foreign to how we naturally are. Separation runs contrary to how Nature works. It is unnatural for life to be so cleaved apart. This denies the constant growth and evolution of reality. It locks life into unending grooves that easily turn into stagnation. It forces a circular system down a straight track.

Humans have been separated from their ancestors, and many stories, legends and myths from our sacred past. We yearn for them because unconsciously we remember them, and how our connection to them used to empower us in the past. Remembering that connection is the beginning in re-creating it in the now.

Every single thing in the Universe is connected. This is a traditional concept of the all-encompassing nature of the Creatrix. We are One. We are facets of the same crystal. Our sparks of consciousness, or souls, are the cells in the body of the Creatrix.

We are taught and forced by society to be separate, but this goes against the Laws of Nature. Separation makes us weak as a species. Nature is based upon dependence. Look at her strong and unified front that she presents to us. Unity is strength. When we are in service to the Creatrix, the lines of separation disappear. We recognize the world around us as a unified ecosystem. Each part is dependent upon the other parts. Natural unity based upon acceptance is the way of the Earth.

What do the Earth and her natural systems teach us? By their harmonic, symbiotic, self-organizing relationships, they show us that we are allied to the whole by the layering upon layering of our cyclical patterns in relation to others. The Earth has so many creatures that work together towards an equal and all-encompassing goal of harmony and interconnectedness.

There is no categorizing and separateness in our bodies, the forest or ocean. These huge systems need no standardization or statistics to function. We have access to exceptional systems of coexistence at our fingertips, in order to mimic them. Nature is a fantastic role model when it comes to functioning systems. She is our greatest teacher.

# loneliness

Really what all the separating, labeling, organizing, sanitizing, compartmentalizing and standardizing does is make for a bunch of lonely humans. We exist in tiny compartment of being because of the separateness of our homes, headphones, families, workspaces and cars. We are cut off from the Earth, our ancestors, from each other, our children and the world around us.

Sure, we have friends and family; but before, we used to have community. We shared much more of our lives. Yet we had a different concept of respecting each other's space, at the same time. We inherently knew to mind our own business. This balance led to happy healthy humans, which is our celebrated past and future.

With all the modern fanfare of communications, the result is desolate humans. We live in boxes, plugged into electrical devices at the end of the day, getting intoxicated, watching fantasy pictures that mesmerize and demoralize us. TV drains our vital life force as it flitters its blue-grey ghosts upon the wall. Walking the streets at ten at night, it is spooky seeing this unnatural grey light flickering through the windows of most homes. It makes our communities like zombie ghost towns.

We have lost our ability to function properly with other humans in a loving and respectable way. People seem to have no manners when it comes to gossip, staring, interfering, minding their own business and keeping their mouths shut. When we lived very closely together, there were rules to our socializing, and for good reason. Rules that were concerned with respecting the personal space of others.  These days that respect is fleeting.

This loneliness produces a huge societal current of sorrow.  Earth has been known as the "Planet of Lost Sorrows" for millennia now.[22]  When you walk on the street and no one is smiling, it is an indication of how lost in loneliness we collectively are.

This epidemic of loneliness is not really talked about.  Everyone feels it on a core level and believes they are powerless to stop it.  It is wise to intentionally reach out to others.  In that sea of sadness lies the solution.  It takes a heart to feel loneliness.  So at least you know you have one.

## political correctness

Standardization is coupled with its bedfellow, the notion of political correctness.  It is the theory that for us to cohabitate with others, we must make our common ground as neutral as possible so as to not offend anyone.  Making everything not offensive is draining the fun, the art and the beauty out of the world.  It places our growth in a strait jacket.

There is no way we cannot offend somebody, sometime.  The world's rules are being made by the prudes and scaredy-cats.  Not offending has become more important than personal expression.  This denial of our natural flow can make expression become stagnant out of disuse.  This can cause us to become numb, which can eventually turn to disease.

Political correctness goes against Nature.  It is trying to make everything fair for everyone.  This just is not the way in life.  Life is not fair because all humans are at different stages of their soul's journey.  We are all experiencing varied levels of lessons.  Some people need the rough, insensitive and callous experiences in order to evolve.

Some people's lessons are totally centred on working through their judgments and misconceptions. They need to be offended to learn not to be offensive. Working through our reactions is how we learn to express love free of conditions, and complete acceptance of reality.

By removing the disagreeable, how are people to learn to ignore it? And by what right do some people have to define what disagreeable is the first place? No one has the right to decide for others how they should feel or express themselves.

The morality of duality is what is keeping this fabricated ethical system of political correctness in place. It squeezes humanity into its little cage of correctness. Humanity is a big ball of rolling fun. It does not want to be crammed into a mold manufactured by a non-corporeal system, with its preconceived artificial notions about what humanity is. Humanity needs no rules or help in defining, expressing and being who it is.

Any kind of deviance from our natural path of development creates discrepancies. These formulate as neurosis in humans. Especially when those limitations are being placed upon us by a heartless, psychotic institution. The system is unknowable and uncontrollable. We as humans will never be able to know the intentions of the system towards us. Therefore, it cannot be trusted.

Political correctness is hindering our development. It tells us to hold back energy that needs to be expressed. This causes lethargy and numbness that leads to disease. It is wise to ask ourselves why the system wants to cause disease in humans. Could it be that it profits from our ill health?

The universal balance that needs to be maintained means that some people are more powerful than others. They do not need life to be sugarcoated to function. Some people are weaker. They are meant to be offended because this enables their personal growth, so they are no longer weak. The truth is the real goal here. Sometimes the truth causes pain. It is wise to suck it up princess, and get on with it.

Why is the system holding society back from its growth and development because of the few? Life is unfair. Life is rough. Life is full of enough bullshit already. By making these absurd rules, it just makes living more difficult. Humans need clarity. The way to achieve this is by calling a spade a spade. Fat means fat, an asshole is an asshole and heartless means heartless.

Political correctness is another name for control. This type of control is deadly because it limits what we can think, feel and speak. It is putting up false walls that are trying to halt the flow of human nature. These kinds of restrictions are exactly what cause discomfort, disharmony and disease. It goes against everything that it means to be human. Not to mention is hypocritical in the face of freedom of speech.

Political correctness is the system telling us how to be. It dictates to us what to say and what to think. It sets the rules for our conduct. The system does not have the right to tell us how to be in any capacity. What we think is our private space. No one has the right to place rules on what goes on in our brains, especially when it is a system of duality that inevitably and subversively is telling us that we are "wrong." "Wrong" is an illusion.

Who creates these rules of etiquette anyway? Why do they not find war offensive? Why do they not find poverty offensive? Why do they not find lying political leaders offensive? Why do they not find police brutality offensive? It seems that political correctness is full of malarkey, because it is not effective at ferreting out the real issues of injustice that exist in society today.

We are taught to be friendly and kind. We are lead to believe to be a good human, we need to be nice and compromise. And yet there are flourishing wars, rampaging corporate greed, and growing numbers of starving families. Maybe if we were firm instead of nice, we could elicit more beneficial change.

Non-discrimination means that we are giving priority to the weak over the strong. This goes against the Laws of Nature. Discrimination follows Nature's way of the alpha leading the pack. It is natural for the powerful to dominate the weak. Why would we want to have people work for us that are unqualified, but fit the quota? This is a wacky concept that promotes more weakness in the population.

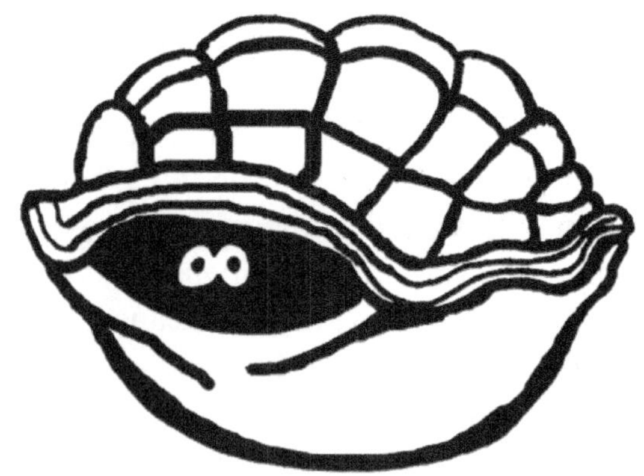

fear

The reality that we live in is inundated with fear.  The media, ambulances, the TV, movies, the police, traffic rules, bullies, social etiquette, jail, disease and laws are crushing in upon us all the time.  The system, the media, the government, the police, society, our peers and our elders are all trying to threaten us in order to dictate our behaviour.

Collectively humans are held in place by tradition.  Tradition is molds of morality and precedence, enforced through fear.  They are constantly telling us to be "good" or else the "bad" man will come to get us.  They constantly threaten us with disease, harm, death, punishment, torture, and hell.

Do you build your life around your fear or do you walk through it?  To be truly human, we need to walk strongly through the fear.  On the other side, we will realize there is nothing to fear.  Fear is an illusion.  It desperately wants to pull us down its slippery slope.

Through the collective unconscious, fear is magnified.  Millions of people share the same fears.  The fear itself is amplified, becomes alive and wants us to join the group so it can suck our vital life force.  We do not have to buy into the hype.  It is only a hallucination and a disjointed dream of the weak.  It is a mere illusory projection, like the one performed by the Wizard of Oz.  The truth is behind the curtain, you just have to walk past your fear to see it.

Unfortunately, it seems to take a terrible tragedy full of fear and pain for humans to react.  It seems we do not learn with love but only with pain.  Will it take the Earth hitting rock bottom before humans collectively wake up and get the nerve to individually and collectively stand up and do something about it?

We have let the fear create rules that we obey.  We do not allow people to have relationships with our children because of the fear of a pedophile.  We do not walk through the night because of fear concerning muggers.  We do not want our women to express themselves fully because of the fear of rapists.  So effectively the pedophiles, muggers and rapers have won.  They dictate to us our actions and we obediently follow their way, instead of the way of love and acceptance.

The system is paralyzing us with fear on purpose.  Funny how this is how it profits - from a population held in the grips of its own self-created nightmare.  Once we understand this concept, it becomes easier to see through the movie set of control.  Pay no mind to its hot air and pomposity.  For like a pimp or a politician, it is all talk and swagger.

When we are afraid, what is occurring is a form of initiation.  Spirit and spiritual people have been purposely hunted and killed for so long now, the fear is natural.  Eventually it becomes easy to disregard it.  It is wise to empower ourselves by walking through the door of fear in any situation. There is a gift awaiting us on the other side.  Take a risk and plunge through. We might surprise ourselves at how much we enjoy it.

We do not have to believe the fear is something coming to get us and harm us.  Instead, regard the situation as our soul wanting us to become more powerful.  We can welcome fearful situations into our lives.  So that we may grow and develop into something we were not before.  Fear is only power waiting to be used.

It is wise to avoid fighting back.  Nothing can harm us, except ourselves.  Just go with it.  Accept it.  Walk through it, head held high.  Be free of fear.  It is the fear itself that creates the scary experience.  Everything that comes our way is in our flow, moving like water.  We can handle it.  The Creatrix will never give us something unless we can handle it.

Fear comes from energetic blockages.  If we undo the pattern that is creating the energetic blockage, the fear will go away.  It is a flag warning that something needs to be changed.  Fear is a constant companion when we are ungrounded.  It is our body's way of telling us that all is not right in its world.

Humans' biological systems are highly, collectively, stressed out. This forces the body to use vital minerals to support the taxed immune system.  This can quickly turn into a mineral deficiency.  Humans are collectively, as a culture, deficient of vital minerals.  This leads to fear and paranoia.  A cultural dip into insufficient nutrients from collectively eating a lot of processed artificial food.

Paranoia is a part of reality.  Fear is a weapon the system intentionally uses in the war on consciousness.[23]  This reality is based on the paranoia of humans.  Do we realize that we live totally collectively paranoid all the time?  It is a symptom of war, control and post-traumatic stress disorder.  It does not have a foundation within the Laws of Nature.

Fear can be related to the issues of security, stability, and the losing of one's possessions.  If we closely scrutinize these fears, we will realize they are illusions.  Security is an illusion.  The Law of Perpetual Transmutation of Energy, a Law of Nature, says change is the only constant of the Earth. Stability has to do with the claiming of our personal power.  Possessions will not go with us when we move from this lifetime to the next.  We are born alone.  We will die alone.  The only thing we will have with us is our soul.  Hence the reason why it is vital to empower our relationship with it now, in the present, while there is still time.

Having fear is normal and healthy.  It keeps our lives in check.  It is the over-abundance of fear that is strangling the human race at this time.  Fear is only ever meant to be a tool, not a tyrant.

You will eventually manifest your fears because the electric current of fear is powerful enough to make itself a physical reality.  Your fear will draw things to you.  It is wise to control yourself and not let yourself become overwhelmed with fear.

So many things in reality are a fearful vortex just waiting to suck you in.  Once they have you, you are fighting the suction of a downward spiral.  Fear is contagious.  Fearful people try to make others afraid as well.  It takes an empowered aura and a strong connection to the Earth to feel the truth in life and to not let the psychodrama of others affect us.

The Goddess resides in the dark forest.  The system with its horror movies, creepers in the dark and rapists lurking behind every bush have tried to spoil our connection to the woods.  The forest is the source of our primal powers.  Ironically, the dark woods is where our salvation truly lies.  It empowers us to walk fearlessly through the night.

Fear is the illusion keeping humans from collectively realizing we live in a garden of Paradise.  It is the most beautiful, luscious, fertile planet in the Universe.  Fear keeps us from walking in our garden.  It is wise to claim dominion of Earth for humanity, instead of giving our power away and letting the inorganic side of life have dominion over us and our home.

## punishment

The system has developed the concept of punishment for those who do not conform.  There are many forms of punishment in our society, from parking tickets to jail.  Jail is torture.  Punishment is not dealing with the causes of the behaviour. Instead, it creates more angst, terror and frustration.

The causes of deviant behaviour lie within the system itself. As a psychotic entity, the system cannot tolerate any criticism of itself. The system profits mightily off the so-called "bad" doings of humans. Ironically, the system itself promotes the behaviour that causes the need for punishment in the first place.

The system is intentionally creating the poverty that drives humans to act irrationally and desperately. The system is intentionally creating the pollution that makes us crazy. The system is intentionally creating the war and the violence to make us weak. The system intentionally creates the mold we fit into in order to be productive within society. It is not people's fault when they are unable to fit the mold.

For millennia now, we humans have been threatened with the loss of our souls and with "hell" for supposed "evil" behaviour. "Hell" is an illusion created by religion to fund itself. Your soul is yours forever. They were lying the whole time. Religions have profited off the threat of hellfire for millennia and shackled humans with its heavy burden.

You can do no "wrong." All actions are simply life lessons. A human can never lose their soul. It is with you for all of your lifetime and lifetimes. Nothing in the Universe can ever sever that connection. You cannot ever sell your soul, either; that is a myth. You can rent it out though. It is wise to realize that all the fear and punishments are empty threats made by an inherently powerless system. That is why it has to steal its power from us humans. Because we are naturally more powerful than it.

**profit**

Profit is "god" of the system.  Speaking of weakness, isn't it interesting how much money the system profits from human addictions to alcohol and tobacco?  Why would our reality be so flooded with heroin and crack unless the system were allowing it and making it happen?  War, sickness, violence are there because the system wants them to be.  It profits immeasurably from them.  Gold fever has taken hold of the population.

The system is busy helping us on one hand with the welfare state, garbage collection and beautiful flowerbeds.  Meanwhile the other hand wages war and massively profits off the debaucherous addictions of millions.  What are the real intentions of the system?

How about a nondenominational system that actually did what they said they were going to do and were held accountable via co-op style leadership?  A system who would strengthen education, health care, but for real?  Instead of a system that lies and forces people to participate in de-evolving behaviour patterns for its own benefit.

Convenience is the lubrication of the god of profit.  Things are convenient instead of beneficial for the health of humans.  All convenient things are is a convenient way to die.  Convenient food contains no nutrition.  Why, and for how long, have humans traded nutrients for convenience?  It is only made so that the corporations may profit.  Sacrificing quality for convenience is unwise.

Preservatives are the foundation of convenience.  All they do is preserve you in an unhealthy state instead of fueling you in a nutritional manner.  Preservatives equal mummies.  They cause cancer because they are so unnatural.

Convenience is to support our fast-paced society.  The only reason things are fast paced is to increase profit.  Society is filled with so many mundane activities, to keep you busy and not focusing on soul because you do not have the time.  We are selling our vital health for the benefit of the few. The Earth is suffering and dying for the profits of a handful of old men.  Seven billion people are slaves to the profit of three hundred elite.

Profit itself is not the problem.  It is the greed for profit that is out of balance with the concept of preserving the natural resources.  Resources are finite and will not support the never-ending hunger of the system forever.  We know we need to take better care of our resources.  Why do we not see and hear about it happening?  Because it threatens the profits of the system.  Native cultures for millions of years were able to work in harmony with the Earth because they only ever took what they needed.

## the system is an illusion

Those are not chocolate chips in your cookies, they are flies! When we look carefully at the system, it is inherently flawed. Under close inspection, it makes no logical sense. It is like an optical illusion. It does not work. That is made apparent by the state of the world today. It functions only upon bravado and intimidation. It is no better than a playground bully.

The system does not have any foundation in the Earth or her Laws of Nature. It is only powerful in the imaginations of humans and only because we have been tricked into believing in its power. The system does not have our well-being at heart. It is wise to avoid trusting it.

The system is totally hypocritical. It says not to kill, but it kills. It says not to cheat, but it cheats. Police are there to regulate the people, but they do illegal things to do this. War does not create peace, it just creates more war. You cannot have a strong foundation for life built upon hypocrisy, treachery and lies. The end result is a corrupt society, which we are neck deep in.

The system is intentionally taking away our natural rights. Its power does not stem from the Laws of Nature. It is not natural. Therefore, it needs to steal ours. The same as any bully. Eventually we will have to face it down. But we, as humans, have the power of the Universe on our side. A fact we seem to forget in all of the crazy hustle and bustle of our reality. There is nothing that can stand up against the bright hard light of the truth.

The system says we live in a democracy. But this is an illusion. We are told we have a choice when we vote for our leaders. Many times the leaders all belong to the same secret societies, which have agendas outside of democracy. Their main agenda is to protect themselves and their wealth, to our misfortune.

It is not a real choice when all we have to choose from are five different crooks. That is no choice at all; it is just more of the same old bullshit. It is the foundation of the system that is corrupt. So any leader we choose will still be the ringleader of a corrupt circus. It is the underlying rules and procedures themselves that are questionable.

These leaders do not properly represent the people. They are mainly rich, white elite old men. They do not have anything in common with their voters. They do not care. The system thinks it can lead us humans, but it cannot. It is like a monk trying to be a DJ for a disco party. It is so separate from us that it has no idea what it is doing. It is completely unrealistic.

The system is like a golf course. Pretty, green rolling hills, quaint ponds, but it is there to serve only the interests of wealthy old men at the exclusion of all others. A golf course all lush and perfect, when the rest of the town is brown and dying from water restrictions because of drought. It is the opposite of wise to have a system so inherently flawed dictating life on Earth.

The traditions that come with the system are false. All of the parades, uniforms and ceremonies have no foundation in the Earth. Rituals of the old school are questionable. We accept them, but they are not assisting us. In fact, they cost so much money that could be put to better use. They just empower the system. Traditional marriage seems more about ownership than partnership.

The system functions by coming between humans and what we endeavour. As soon as this middle step is created, the real power, the juice and current is weakened, drained, and misguided by the interference. This interference disrupts the original output. It then turns it into something it was not, creating something that is artificial and not wholesome.

The system has a tendency of wrapping its most vicious things in pretty packaging. It is wise to be untrusting of things that are pink, sparkly, happy, fun and tasty. Candy looks colourful and happy, but it is poison. Fried food smells great, but it is poisonous. Pharmaceutical drugs say they will help us, but are poisonous. Walt Disney is so warm and special, except it has stolen our heritage and tweaked it to be nonsensical.

The system makes things light and fluffy on purpose in order to distract us from the true nature of reality. Things are shitty, let's make everything happy. La la la. The system invests incredibly huge amounts of money into things that are meant to distract us like the Olympics. This money could be used in such better ways. This is insulting and degrading to the human population.

## intentional suppression

The system is creating this desperately horrible way of collective life that we reside in today on purpose. Can you see the pattern? If you looked at just one issue, you may have passed it by. Put a few of these issues together and maybe it is just a coincidence. But when you analyze these issues together, the straight systems versus the spiral nature of the Earth, you realize this is no mere coincidence. The system is suppressing humans on purpose.

The system wants there to be war and poverty because this is how it profits. It creates stress on purpose like horns on cars, sirens and obnoxiously loud noises. It creates horrible jobs, backbreaking labour and never-ending work schedules to fuel it. What is the point other than to stress out humans because of its macho ego needs to be fulfilled by the peasantry?

Our reality is dictated to us by the stores around us. It is near impossible to have a healthy lifestyle if the only choices are fast-food restaurants, coffee shops, liquor stores, and convenience stores. It takes awareness and dedication to have only real food in your life. When you do, the rest can quickly fade away.

For example, young people at 11 o'clock at night are looking to have some fun. What are their options? Bars. That is it, there is nothing more offered. There are no places of games, swimming, aerobics, trapezes, miniature golf, pool, or a million other fun things to do at night other than getting wasted.

Do not be naïve and think that those five years of accepted and prescribed partying after high school do not damage our brains and livers in those incredibly formative years. It sets up our entire futures with little or no sparkle left in us.

Like cattle to the slaughter. The irony is that the cattle willingly put their heads on the chopping block. They are too ignorant to know better. Then of course, the old cliché about having to settle down and get married is the only solution. We do not have enough sparkle left to get on with being self-blissful for the rest of our lives. It is easier to give up on our dreams and simply conform.

The system pushes the young people to have addictions because it cuts off access to anything else. It knows if it pushes hard at a young age, it will be able to effectively create a generation of drug-addicted alcoholics. It is not by chance that sugar is so readily available for children.

This is an intentional situation created by the system to make children weak at a young age. To ensure a lifetime of addiction. It is no coincidence how much the system profits from the addictions of its citizens.

Notice the system slipping in between parents and their children. It creates such addictive distractions that make parenting very difficult. It threatens parents with jail if they refuse to comply with putting their children on behaviour modifying drugs. It makes it illegal to punish children as parents choose.

Notice the system trying to disable our connection to the Earth. Notice it polluting our blood. Notice the system is so contemptuous towards soul, energy, and spirit because it knows that these are the only things that will save us. There is a huge slander campaign against anything spiritual because it is such a threat to the system.

The system is intentionally slaughtering Nature because Nature is where the solutions lie. It is where our power resides and the system is desperately trying to cut us off from our power source. It knows we will blow that system away like dust once we have collectively reconnected with Nature.

Look how the system ridicules and puts down anything and anyone who challenges its authority. They call it terrorism or treason if you speak out against it. The system has things set up so as to weed out the natural sources of sparkle. Then they try to sell the sparkle back to us in diluted, processed and heavily taxed forms.

The system focuses upon the daytime, as it the human's weakest time. The daytime is the grossest expression of the human. In fact, our subtle body, our heart and our feelings all function best at night. And then they create a huge fear campaign against the nighttime. Nighttime is our sacred time. The woods are our sacred ground. The system intentionally tries to cut us off from our inherent power at every turn.

The system makes broken families on purpose. It has made gathering large groups difficult and illegal in some cases. It is waging a war on the consciousness of the human. It does everything it can to bring us enough stress to make us crazy. But not too much that we cannot go to work.

The system wants us to be damaged because then it knows we will not fight it. It wants us to be so busy with our own survival, so we do not care about anything else. It does everything it can to distract us from the true nature of society.

The Earth naturally provides for us. For millennia we survived from her bountiful harvests. Now the Earth is incredibly manipulated and controlled. The resources are being hoarded by a minute few. This makes humans struggle to feed themselves, and poverty is rampant. This minute few intentionally create scarcity, poverty and overpopulation in order to profit.

Overpopulation is the crux of the problem. It is caused by the overuse of pharmaceuticals. It is causing the balance to be fucked. It creates the food equation to be skewed. It creates rampant disease and poverty. It hurts the Earth because of overconsumption of resources. It creates pollution. It is the root of violence and corruption.

The system is intentionally creating overpopulation in order to profit off the scarcity vibe and the upset feelings it causes in humans. We are like grasshoppers. The system knows that eventually our wings will brush one to many times with other grasshoppers, and we will turn into a plague of angry locusts. It keeps us on the brink of being angry locusts in order to profit off our agitation. Because to destroy everything around us would affect its profits.

It is wise for humans to focus upon their connection to the Earth. We can return to her and reconnect with her. In this way we can have all our needs met without the use of the system's middlemen. Right now, because of the system's interference there is a terrible cacophony, when really the Earth is a beautiful orchestra. It is a choice as to what you play on your soul radio.

## solutions

We as humans are naturally chaotic. That is not to say we do not excel with boundaries. It is not to say that the concept of the system itself is flawed. It is that the source of the system as it is now does not work efficiently for humans. It is inherently out of order. There must be a way that the system can be altered to consciously put the whole forward, instead of itself.

The system is not "bad" or "evil." It is just not in balance with humans or our planet. It comes from the top down, instead of the bottom up.[24] The solution is not fighting the system. Not in its destruction or annihilation. The solution is in bringing balance to the system, by bringing balance between technology and spirituality. The system needs to be expanded to match our expanding consciousness. There must be a way. We just have to find it. It is wise to not give up searching.

The goal should be the merging of the two worlds by bringing together the city dweller with the forest dweller, technology with spirit, woman with man, the rich with the poor. We need to correct the system's straight lines with spirals. It seems difficult, but only because we have not given it enough thought.

There has to be a way for technology and Earth medicine to get along, a way to find the happy medium. There has to be a way for Earth ways and star ways to combine and form something new. A way that benefits all, serves the whole, and does not damage the Earth. The system is flawed. But surely, there is a way to restructure it to make it work. Combine the Laws of Nature with technology to come up with new systems.

Being free of the system's tyrannical ways is a matter of collectively raising our vibration.  The war vibration is a shallow vibration, mean and cruel, of fighting and first chakra aggression.  We can raise our vibration systematically with dedication and awareness and some research.  When we collectively raise our vibration, we live in another place.  It is literally another world.  The shallow vibrations cannot even see us.

Ten is the illusion the system tries to get us believe in.

Eleven is the sacred space that is created when two spirit people get together and co-create.  A Sunspot cycle happens every 11 years.

## sacred mirror

The sacred mirror is an incredibly beautiful and intimate experience. A woman will see her inner God reflected in her partner. A man will see his inner Goddess reflected in his partner. That is why you draw certain partners to you, so you can better see yourself.

You project your feelings outwards onto others. They are then reflected back to you so you may learn about yourself. This is the sacred mirror. You learn about yourself a great deal by asking the question, "Am I?" The sacred mirror answers, "I am," as you view your reflection. Having a partner can be an intense experience of the sacred mirror. Your lessons are in your face. You cannot avoid them.

This process forces you to deal with who you are by experiencing it firsthand. You then have the decision to accept it or change it. Sometimes it is gratifying to look in the sacred mirror and receive the acceptance you have to give yourself. Sometimes it is painful to look in the scared mirror and accept the truth about who you are.

When you do not like someone, you are projecting what you do not like about yourself onto the other person. They are reflecting it back to you. This is why you become irritated. These are the lessons you need to work on within yourself. It is a time to change and move through the things that aggravate you about yourself.

It is wise to process these feelings by feeling them fully. Then consciously release them if they no longer are serving your best interests. You can say something like, "I acknowledge I feel this way. I know it is a part of myself I am needing to experience. I consciously release it."

Why is that person bothering you? What are the things that are triggering your adverse reaction towards them? What is the issue you are projecting onto that person? It is wise to look at it, explore it and be open to receiving the lesson. All you have to do is learn the lesson and eventually the upset will pass. When something is irritating you about another person, those feelings are coming up because they want to be moved through. It is a beneficial thing.

When you point a finger, three are pointing back at you, literally. It is wise to show gratitude to the people who stir up the unfavourable emotions in you. They are assisting you with your personal evolution. This is an example of the paradox of spirit. It is wise to thank the people who piss you off.

The scared mirror is what will spiritually advance you the most in life. Being alone you do not learn the lessons at the pace you do when you interact with others. It is wise to learn to accept what you see in the sacred mirror. Take it in and learn from it. It is truly trying to assist you in bettering yourself by raising your vibration.

The sacred mirror assists you in defining who you are. The reflections you give others aid them in defining themselves. The sacred mirror naturally happens with everyone you meet. It compels you to see the truth about yourself, instead of just guessing. It forces you to push through the difficult parts of your personality. Accepting things about yourself you probably did not know even existed. It exposes your under lovelies.

The sacred mirror shows you that you are not alone. You are connected through this game of projection and reflection. You need others to assist in knowing yourself. You cannot avoid the sacred mirror and its powerful echoes because it is you. The way to be free is to go through them, instead of trying to run away from them.

It does not matter who your partner is. The same lessons will still come up whoever you are with. You can try to change partners a hundred times but you cannot escape your lessons. Your issues are your shadow because they are you. You like to blame others for the weirdness in your life, but it is not case. You yourself create it all. Once you accept this, life will flow more smoothly.

The scared mirror enables you to go deeply into your character to break down any inconsistencies. You just have to be open to receiving the criticism. Change the parts you do not like about yourself. It is wise to try to control what you project onto others so that you can avoid the pain of repugnant reflections bouncing back to you. The best way to do this is by remaining equanimous to others. It is wise to learn to practice complete acceptance.

The sacred mirror is the most efficient way to process your unconsciously held beliefs. Before you know it, they will rise up and demand to be dealt with. Your partner is someone in your face who will hold you accountable for what you do. It is not so easy to sweep it under the carpet.

It is wise to realize that every single emotional issue you have with another person is you projecting your emotional landscape onto them. They are reflecting it back to you. It is not wise to hold others responsible for your shit. It is you, not them.

When you can take full responsibility for the intense projections that you do, it will become more obvious to you. You will probably do it less because quite frankly, it can be embarrassing. There are easier ways to deal with your baggage than by getting others involved.

The sacred mirror assists with your personal alchemy. Together, the two of you have to live the consequences of each other's alchemical choices. The energy ratchets back and forth, deeply affecting your partner. If you drink, then it is like they are drinking, even if they are not. If you pollute your body, it affects the other. You have to deal with the other's alchemical choices because you have to kiss them and smell their breath and body odour.

It is wise to reach together towards the highest vibrational alchemical consumption. Sometimes that means nagging and bitching. Especially if your partner has farted up the room so bad, you cannot breathe because of their not-so-wise alchemical choices.

The sacred mirror has instant karma. Whatever you put out quickly comes back. Not as much fun when you are on the receiving end of your own shit. Then you tend to work through it quickly because you do not want to have to deal with it anymore.

As a society, we tend to throw our poopy underwear on people when we should be quietly doing our laundry by ourselves. We are gossiping about their dirty laundry before realizing that we are still wearing shit stains on our own pants. It is wise to keep our mouths shut and stop judging others.

When people are irritating you, it is wise to remember that you have been there yourself. There are times where you have been a dick, a liar, an abuser, and a schmuck. You have been reincarnated, you can put yourself in their shoes. It does not mean you have to care or concern yourself about their reality. But you can still accept them. This does not mean that you have to participate with them, though.

You do not help yourself or others by isolating yourself. The world needs you and you need  it.  It takes more courage to share yourself than to live in a secluded bubble.  Even if that bubble is paradise.  Ideas need to circulate in order to be affective.  The world will evolve more quickly if ideas are shared instead of being hoarded in hidden pockets.

At the same time the sacred mirror can sometimes be too much.  You do not have to punish yourself needlessly.  If you end up living in a warzone with your partner, you can always stop the crap from flying by walking away.  You can always find a new partner.  This time you will be that much closer to knowing what you want and need out of life and a consort.

# triggers

The sacred mirror can produce a lot of energy.  It ratchets back and forth between partners.  Sometimes it can get nuclear.  This teaches you how to be patient, not to name-call, to listen and to accept.  Sometimes these are not easy lessons.

Soon you will learn what each other's triggers are.  A trigger is a word or phrase that will set your partner off.  For whatever reason these words have a heavy charge and you partner is still working it out.  In a way, it becomes easier as the people in your life assist you in working through these reactionary phases until you are free of them.  Slowly you can learn to respect each other by not aggravating the other's triggers.

Love free of conditions translates as assisting each other to move through the heavier energetic content.  It means not reacting when someone has a burst of energy because they were triggered.  When the other person also does not react, it allows the person to move through the outburst without attaching to it. So they may be free of the pattern.

We all have triggers. Who knows how or why they develop. They just are. There are certain things that set you off. A true partner will not antagonize the other by intentionally pushing their buttons. They know it will not be long before their partner will turn around and push back. Eventually a truce can be called. This can allow emotional intimacy to develop at a deep level.

# i am a mirror

Remaining equanimous becomes easier when you realize that people are projecting their beliefs and judgments about themselves onto you. Their stuff has nothing to do with you. The way they perceive you is their judgment. It is not you. You are not responsible or held accountable to the byproducts of the imaginations of others.

If you can understand that you are a mirror, you will realize that you are not responsible for what people believe about you. It is quite freeing. You become unburdened of the projections and judgments of others. You no longer take their shit so personally.

When you realize you are a mirror it makes it easy to slough off any weird notions given to you by others. It is not you. It was never yours to carry in the first place. You do not have to accept it.

Holding on to the judgments and expectations of others is a fast way to age yourself dramatically. It is wise to be free of accepting and trying to fulfill the projections of others. Allow them to pass through you, but do not attach to them as they go about their merry way.

You can say right outloud, "Hey. You are projecting your beliefs onto me. I do not resonate with what you are saying. I am a mirror. Maybe you need to think about what you are putting out right now." Eventually you will not need to say anything at all.

You have to accept the mirror. Accept the fact that people are really just seeing themselves when they look at you. What is reflected back to them is a lesson they need to learn about themselves. It has nothing to do with you, unless you want it to. This works wonders in conflictual situations. It puts all of the name calling and bullshit right where it belongs. Back in the lap of the giver.

# expectations

If you get upset with others, it is because you are projecting expectations upon them. You are disappointed when they did not fulfill your expectations. Your projections are quickly reflected back to you. It is wise to accept people as they are. Instead of trying to make them into something that fits your definition of them.

"If you really loved me you would..." These kinds of expectations are conditions. These conditions limit your ability to experience the limitless nature of reality. You create shallow vibrational relationships when conditions are the foundation. Society has been functioning with conditions as a standard form of currency. There are more efficient and effective ways to establish heartfelt relationships between humans.

Living your True Will is the most important thing you can do in your life. By compromising, you are limiting your ability to express yourself fully. Compromising also turns your power down. It is trying to control and funnel your Will down artificial tracks. Instead, you can let it blossom and grow like wildflowers.

Putting someone before your True Will is not love. True love is allowing and desiring your partner to fully express their True Will. This is love free of conditions.

When you truly love another, you want them to be in a state of bliss. You value their attainment of bliss over your own connection to them and your own quest for satisfaction. You wish for them to propel through their evolution. You want only the best for them, regardless of how these things affect you personally. You can make others shine like stars by putting them on pedestals of bliss and acceptance.

Why apologize, unless you want to? You are on a soul path and are learning your lessons. Learn your lesson and do not do it again. True practitioners of love free of conditions do not have expectations. So they would never get offended in the first place by anything that is said to them. They accept the lessons of others. They do not get insulted over others learning their lessons. There is nothing to apologize for. Apologizing creates karma for no reason. It is more respectful to discuss it with your friend and then just let it pass.

Society pushes expectations and says they are real. Political correctness dictates a standard set of expectations and labels them life. In this way the system can cage the wildness of humans and ultimately profits of their typical docile behaviour. Expectations is the religion of the system.

Breaking out of the mold of expectations aligns you with your true self. Maybe that is not for everyone because it is just too darn exciting.

Hints for the wise

- ◆ It is wise to find friends and partners who work with you and your triggers, instead of those who enjoy pulling them.

- ◆ It is wise to find people you resonate with and support them. Find stores with owners who share some of your beliefs and give your money to them. If a place makes you feel uncomfortable, do not support it. You have the choice where to put your energy, so choose wisely.

- ◆ It is a great idea to find your local spirit store. Here you can not only get supplies, but you can meet others like yourself.

- ◆ It is healthy to find friends who are interested in the same subjects as you. There are spirit meet-up groups all over. You can find them online.

◆◆◆

Eleven represents the power of spirit people co-creating together.

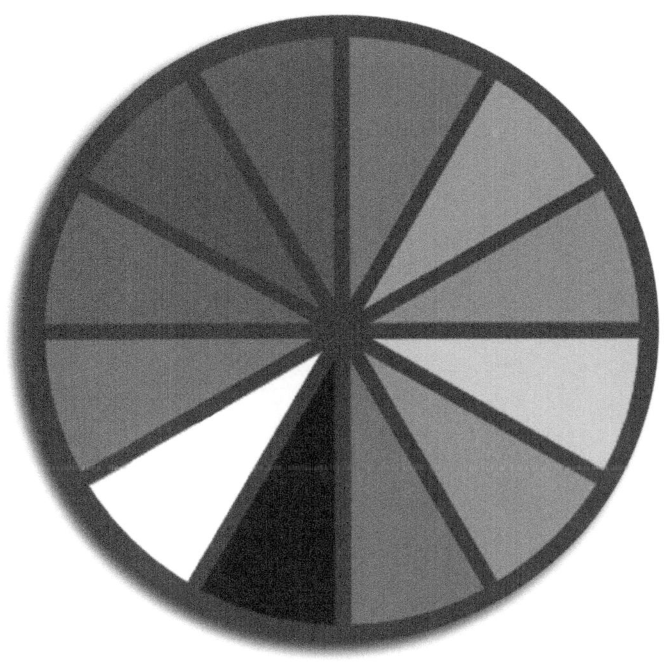



Twelve is completion. A complete cycle: energy, to idea, to manifestation, to completion.  A circle complete, a situation resolved, the aspirations of humans recognized, an end to the journey of the soul, and an immortal human.  There are 12 pairs of ribs in the human body, 12 human cranial nerves, 12 priestesses to make a Merlin, 12 faces of the dodecahedron platonic solid, and it takes 12 sperm to launch the 13th sperm to make a baby.

# Faeries

Elements are born in the heart of a dying star. Each has its own chemical composition and colour. They never die, but instead are recycled. The elements inside of you were once a part of stars, trees, animals, rocks, or maybe even Mozart.

Elements have a spirit or a soul, just like all the creatures of the Earth. The spirits of the elements are called Elementals. Another name for Elementals is Faeries. Faeries are the messengers of the Goddess. Just as Angels are the messengers of God. They are neutral and do not have gender the same way humans do. They are quite chaotic.

Faeries are the Guardians of the Earth, including her streams, birds, minerals, caves, plants, fields, rivers, animals, pools, mountains, insects, and trees. Faeries are the Guardians of all the forces upon Earth, including powers such as gases and energies from radioactive minerals and the phenomenal forces of the Earth herself like volcanoes, wind, tornados, earthquakes, magma, tsunamis, waves, tectonics, and geysers.

Different landscapes have different species of Faerie. There are as many different kinds of Faeries as there are animals. More in fact, each well suited to their task. They tend to group in Clans based upon their elemental content.

There are natural ranks within the Faerie world. They are fractal, not hierarchal. Big difference. They have their own biodiversity and authority. Some Faeries are little happy ladies that fly around your head. This is truly a small percentage of who they are.

The Fey have their own form of governance. It is alpha-based. They have an Empress and Emperor. There are many species of Faerie within their ranks. Like Angels, Faeries have different clans who are headed by an alpha.

At any given time, there is a head of the clan, like an alpha dwarf who is the role model for all dwarves. The King of the Dwarves, he is called. This theory is the structural framework for all of Faerie Land.

It is wise to be very careful in any dealings you have with the Fey. They are incredibly powerful. They are uncontaminated expressions of Mother Earth. Pure sources of primordial energy. Not to be fucked with. They are shy because they can kill humans so easily. They tend to stay away because they are not interested in incurring karma if they accidently harm or kill you.

Faeries do not understand war and murder. They do not understand this concept of fighting. They have been greatly affected by the way humans live. They used to be our friends. Now they do not trust us. Obviously, it is because humans ceaselessly kill them and destroy their homes. It is no wonder they do not like most humans.

The loss of their natural habitat sucks. They do not flourish in cities, as Nature has been so squeezed out, there is nowhere left for them to live. No wonder not many people believe in them anymore.

Faeries are real and alive. They are sparkle masters. They live in a dimension just a heartbeat away from you. They can make themselves known in your realm and engage with you, but only if they trust you. To earn their trust you can leave them offerings of honey and milk, traditionally on Midsummer's Eve. They like tiny presents too.

Faeries are spirited so it is wise not cross them. They are quite tricky. They will try to jest with you, if they can. Faeries like to take things and hide them. If you are fortunate, they will turn up somewhere else later. They enjoy all sorts of tricks. Odds are, if there was some mischief, it was they. Just because you cannot see them does not mean they cannot see you. They truly enjoy fucking with you.

Finding Faeries and becoming their friend is a personal experience. It is one that takes belief, diligence and most importantly an open heart. It is the ultimate reward in spiritual belief when you make their acquaintance.[25] Faeries are quick movers. When you see them, they freeze. You know sometimes when you see flutters and movement at the periphery of your vision? Yes, that can be they.

Faeries are elements. They are in everything because they are everything; as elements are the building blocks of all life. They do not have to obey the same Laws of Nature as you do. They can hide in things and you would never know that they were there. Keep a lookout for eyes because that is how you will first identify them. If you are fortunate, they will blink.

You can find your way to Faerie Land. It is possible. It is the most beautiful and terrible place you will ever go. Oh, the dancing and the roaring death. The sugar icing, sweet action, and the lacey skeletons. It is a grotesque masquerade with giants, rushing time, wee folk, dire smells, curly shoes, electric colours and haunting sounds.

The most incredibly beautiful people you have seen in your life, dressed up to the nines, exuding passion and wicked fancifulness. There is dancing to the most intricate and delicately thundering music you have ever heard. The most lurid and horribly disgusting visions twirling as if in a kaleidoscope. It is wise not to drink what they offer. Best of luck finding your way home again.

Elements, and therefore, elementals never die. They are only ever recycled. They are incredibly old. The Fey carry vast, ancient knowledge because their memory extends to the Earth's beginnings. They are a heartbeat away from the spirit of the Goddess. They live inside of her body. So, they are very powerful.

They still share in the transpersonal memory of the planet. They are intimately connected to her workings. They will share with you, usually for a fair price. They will give you exactly what you give them. You can make faerie offerings a part of your normal life. For example, your daily walks can be a great time to reach out to them. They will appreciate anything you have to give.

The Faeries and Elementals think the people with too many artificial particles in their veins are aliens. They do not like the aliens because they believe they are responsible for the slaughtering of the Earth. Your relationship with the elements is the majority of your reality. To discount the Elementals is unwise.

Just because you are dumb enough not to believe in faeries, does not mean they do not exist. They believe in an eye-for-an eye kind of relationship. Most humans are mean to them. They want you to know they are fed up.

# Angels

As Faeries are the Elemental expressions here upon the Earth, Angels are the elemental expressions within the Universe. Angels are the messengers of God. Faeries are the messengers of Goddess. They are the same thing, just equal and opposite expressions of each other.

There are three original Angel families, each headed by an alpha. They are the Firstborns. They each have their own role, skills and duties. There are many Angels families that were born after the three Firstborns.

There is Sariel. She is the Angel of Death. She is the Firstborn. She is in service to the Dark Goddess. Her minions are the Grim Reapers or Death Walkers, who lubricate the process of soul transportation upon birth and death here in the Universe. The only ones who can traverse between the Void and the Universe without losing their minds.

Grim Reapers are in service to the 13th dimension, the Void and its Mistress, the Lady of Death, the Dark Goddess. This means they are in service to Sacred Law, the primary code of ethics inherent in all spiralized creatures of the Universe. They represent Universal Justice.

Their department has been humourously nicknamed "Creatrix Corrections," or more simply, "Corrections." Their sparkle power instantly brings anything in their periphery into alignment with Sacred Law. Exorcising everything else back into the Void where the energy will be recycled and made use of in another way.

There is Ophan. He is an armed, militant spiritual warrior. His minions are birds, cats, wolf/dogs, horses, dolphins and whale angels.  He is in service to the Goddess and is her champion throughout the Universe.

There is Seraph.  He is the caretaker of the realm of God.  He is the personal assistant and tends to the administration of God's needs.  His minions are large plasma fields that are his messengers and reptiles, including wyverns, serpents and the like.[26]

Angel families are currents of energy.  They can be called upon to assist in the flow of life.  It is wise to research them and find which resonate with you.  You can and make an effort to bring them into your life.  Their energy is sublime and limitless.  It can empower you in unforetold ways.

# Faeries Tales are true

You have a rich legacy available to you through Faerie Tales.  They, along with myths and legends, are your true human sacred heritage.  Humans come from a world of magick.  You still live there, though some may not know it.

Faerie Tales tell you of your hidden powers.  You as a human are incredibly powerful.  Your heart and its ability to feel and to offer love free of conditions makes you special.  You have spiritual powers that are not found anywhere else in the Universe.  Your home is a majestic garden, like no other.

The legends of old speak of your abilities to know things and to see things.  They teach you of your gift of feeling the pea under those twenty mattresses and your ability to fly.  They tell you of the realm of the Faeries where you would dance the night away.  They describe the tunnels that went deep into the Earth, full of gold and gems, always guarded by the Dragon. They warn you of the sorceress as she reaches out to grab you, trying to throw you in the oven.

These fables speak of a world before time, buses, rent and bosses. In Earth herstory there has been such a time. A time with no war, religion or money. You accept Faerie Tales so easily in your life because you feel their validity. You unconsciously long for them again. Your ability to dream and imagine such realms makes it possible for them to exist. If you can conceive of them, you can make them real.

Faerie Tales live in the collective unconscious where they are alive and full of vigour. They fill you with razzle dazzle. Their symbols unfold outwards, beyond time and culture. They reach beyond your logical mind, reaching deep into your heart, where the magick still lies.

Faerie Tales guide you through your unconscious. They offer inspiration. They are full of truth. It is a world where an eye for an eye is law. You instantly feel the consequences of your actions. They still persevere after such a long time because of the power of their truth. In many cultures, you will find the exact same stories because they bypass cultural barriers. They are human stories. You will even find them in the modern era, in the form of urban legends.

Faerie Tales show you how reality is able to bend. This bending is the true nature of this world. There is no spoon. You have been convinced otherwise but that does not make it any less true. A sleeping spell has been put upon the spirit children of the world. *The Trinity Theory* is the freeing kiss!

elementals cause cancer

Elementals are causing cancer on purpose.  For the first time in a long time, there are hazardous elements out of the ground and in your life. Dangerous elements that are not supposed to be anywhere near humans, or the light of Father Sun.  Ill health can come from exposure to these elements. Addiction is based on addiction to elements in the bloodstream.

Society uses many elements in a de-evolving way, and without gratitude.  This causes upset in the energetic world, much to our misfortune. Time to wake up to the fact these elements are alive.  Humans are incurring some heavy karma with them by being so ignorant towards their existence.

Our bodies are made up of elements.  But not all of these elements are meant to be in our body.  The world today uses elements such as lithium, uranium, plutonium, radium, cadmium, and francium in technology that affects people.  Today there can be found toxic elements in unsuspecting places.  This is directly affecting the vital health of humans in a negative way.

Because Faeries are the spirits of elements, they are energetically attached to them.  When you consume toxic elements, the Faeries get stuck in your body.  They get upset because they want to go home.  They cannot because you are imprisoning them.  Unfortunately, when they are stuck in the human body, much to the human's poor fortune, they get angry.  An angry Faerie is a Demon.  So the Demon encourages destructive behaviour because it wants you to get sick and die, for a specific reason.

An Elemental, Faerie or Demon, vibrates at intensely high rates, especially radioactive ones.  When they are angry, they vibrate even more. These intense vibrations can cause the cells they are lodged in to mutate and turn cancerous.  They know this.  They are doing it on purpose.

They know if you as a human get cancer, you will die.  Then they will be able to return home as the elements are released back into the ground upon your death.  Our use of radioactive elements and the rising rates of cancer is intimately connected.

As a human, you can connect with the Elementals inside of you. You can become aware of them.  You can work with them, instead of against them.  Faeries are the spirits of elements.  They naturally live in the ground. Not all of these Elementals should be in your body.  You can assist them in returning home, because they are energy.  This energy can be escorted from your body, back into the ground to be received by Mother Earth.[27]  This will assist in the elements being expelled from your body.  This reduces your risk of getting cancer and disease.

You can assist them and yourself, by aiding them in their return voyage. The way to do this transport is by opening a Creatrix Vortex in the ground. It will act like a wyrmhole taking the Faeries down into the ground and back to the elemental realm. This vortex works because of the natural flow of energy from God above to Goddess below. It is important to use the Creatrix Vortex because otherwise the parasitic entities can and will jump to the next closest person.

Sit and relax. With your hand, energetically open a hole in the ground at your feet. To open this vortex, you use a widdershins (counterclockwise) motion. Lift off the lid.

Ask politely, "Are there are any Elementals inside of me that would like to return home?" Wait. Usually you will feel your body tingle or get warm.

Then go to the place on your body that is tingly. Pull the Elementals away, dropping them into your vortex. When you feel like they are gone, put the lid back on. Close the vortex in the ground with a clockwise motion. Rub your body where they have left. Thank the Elementals for leaving your body. Thank the Goddess for receiving them.

So many of the problems the humans of the world are facing right now are based upon relations with the elements and their spirits, in the form of Elementals or Faeries. If you do not honour your relationship with them, they will kill you.

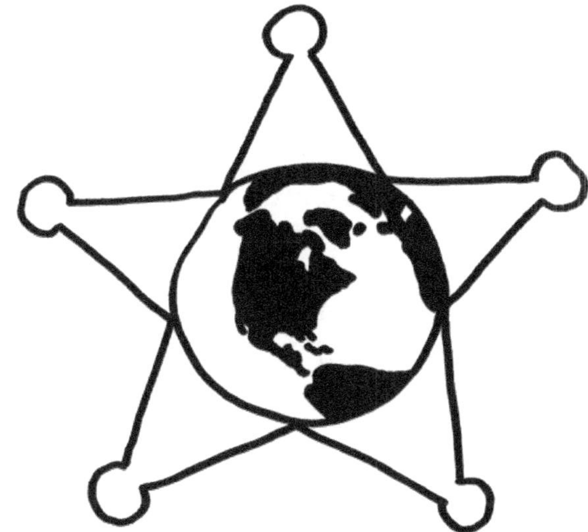

## Guardians of Earth

The Earth is the heart chakra of the Universe.[28] Unfortunately, she is broken and off kilter. That is why war has raged here for so long here. Different groups have thought they could somehow claim dominion of Earth and possess her in her time of distress. They have forgotten the fact that she is an independent entity unto herself. Her memory is much longer than theirs is.

Humans are awakening to the fact that they are the rightful Guardians of Earth. As a Guardian of Earth, you are a protector and lover of the planet and all of her creatures. No one can make you an Earth Guardian because you already are one. You are the one who has to bestow the honour upon yourself. No one can do it for you.

Planet Earth is a special place. It has an amazing wealth of information. The perfection of the natural interconnected systems of Earth have so much to teach you about how beings can function in harmony with each other.

There is absolutely no need for anyone to be hungry, or cold, without shelter, running water and clean air. As Guardians of Earth, you are a protector of the planet and all of her resources. A false, controlling entity has bullied you into believing otherwise.

The Earth and her natural systems are a vast ecosystem that is interconnected and dependent upon one another for survival. As a Guardian of Earth, you become aware of your own connection to this gigantic web of pulsing life. You realize your own intense role you play in its orchestra of amazingness.

A Guardian of Earth inherently understands caretaking is a gift, not a right. You do not own the Earth. You are not its overlord. But instead, you are its student. There is so much to be learned if you allow the flow of Nature to captivate you with its intricate lustre and beauty. Allow it to charm you with its brazen intelligence and powerful fearlessness.

The web of life is incredibly intricate and each single part is vital for the whole. If one aspect is damaged, then the whole does not function properly. When you fully realize this, you will know in your heart we are all one. When you truly feel this, it becomes impossible to harm the things around you. For why would you want to hurt yourself?

Knowing and acting upon being One, in service to the Earth, is your natural way of being. It is your sacred heritage and your immediate future to be fully connected with the elements, animals, stars, plants and wisdom of the ancient ones, once again. The Universe uses all of its immense power to constantly re-balance itself. You have all of its support behind you, propelling you towards this vital future.

Can you believe in living a life free of all limits?  Can you perceive yourself as fully satisfied?  Can you imagine the world  free of pain?  Can you envision all of your dreams fulfilled?

The human heart is the most powerful thing in Universe.  What does not work will eventually fall away.  Humans are meant to be happy, free, nomadic, wild, dancing and with a strong connection to the heartbeat of the Creatrix.

# Hints for the wise

♦   We can create self-sufficient, sustainable homes.

♦   We can create systems that cleanse and purify the water and make it available to all.

♦   We can create our own unlimited power, with alternative power sources like windmills, geothermal energy and water wheels.

♦   We can create a place to have menstruation rites, like the blood rites of the past.

♦   We can create a world with enough healthy food for all.

♦   We can create a world free of military and weapons.

♦   We can create affective ways to dispose of garbage and toxic waste.

- ◆ We can create co-ops that are healthy and work where we need authority that makes decisions.

- ◆ We can create focus on bringing balance to the systems of the Earth.

- ◆ We can create a world where people are accepted as they are in all facets of society.

These stories are true: of the Holy Grail, elves, gnomes, phoenix, satyrs, centaurs, tree people, sprites, trolls, dwarves, Faeries, ghouls, Demons, giants, Arthur, hermits in caves, will-o-wisps, ogres, fay, wee folk, brownies, dryads, selkies, hobgoblin, leprechauns, giants, golden geese, jinns, and siths, for example.

◆◆◆

Twelve represents completion.  It presents truth, which always will shine like a star.  If you could have a wish for the future, what would it be?  Who would you be?  It is wise to define it now, so it is easy to step into when the time comes.  Now it is the time of harmony of the world's orchestra. Welcome back home!

Thirteen is the unexplainable mystery of life.  There are 13 moon bloods
for women per Sun cycle.

## Sacred Law

In the days of old, and for those who still remember, Sacred Law is the law of the land. It is an inherent code of ethics. In the past, it was instinctively honoured by all humans. It is the way of the land. Because it works, it is not something even talked about. It just is. It is definitely not something you are forced to do, under duress, like the system today does to its citizens.

Sacred Law is not something written down. It is passed on through oral tradition. As are all its stories, lessons and ways. This code is integral to being a living creature here on Earth. It is inside of you.

Sacred Law is incredibly simple. This code is based on respect. It is the way of acceptance. It is a code in service to the Earth and all of her creatures. It is a code where the individual is in service to the whole. Because there is this one code, there is no need for rules or laws.

The ones to carry Sacred Law are the Crones, the elder females of the clan. The Crone represents the Dark Goddess, one aspect of the triple-deity, the Goddess. The Crone as the giver of Sacred Law is the wise one, midwife, healer, doctor, counselor, mayor, judge, jury, police and executioner.

It is understood that she is closest to the Goddess. Therefore, she is considered the voice of the Goddess. She has reached the point in her life when the two hemispheres of her brain have merged and work together as one. She has also stopped bleeding, so she is androgynous. In other words, she is neutral and incredibly wise. This is how you know her counsel is sound, unlike any other group of people. She will always be in service to the whole and never to herself.

Having the Crone as the leader follows the Laws of Nature. The Mother lion is always in service to protecting pride. She puts herself aside to feed and take care of all the members, whereas father lion is in service to his position in the pride and his own stability. Father lions make terrible leaders because they are only in service to their ego, concerned with fucking and fighting.

Women will always make better decisions for the whole because they inherently protect the children, whereas men will protect their position first. This follows the Laws of Nature. You see the end result of men being the leaders in today's world as it is full of war, violence and aggression. All bedfellows of men, not of women. Women would never choose this way if given the chance. Society needs to be men and women working together in harmony. Balance between the sexes.

If someone is outside of being in service to the whole, Sacred Law says they are banished from the community, which means death of the worst kind. The Crone speaks the word and you are gone. It is known as the "Curse of the Crone."[29] The community turns their back on you. Alone in a savage world, you will not survive long.

The Crone is in charge of delivering babies. She uses her skills to bring new life into the world. She tends the wounded and the sick. She presides over the deathbed of those close to death. If they are suffering, her word grants relief. Her herbs will free them from their physical shell and the pain it is causing them.

The Crone traditionally represents this pivotal point between life and death. She is given this immense task of holding the power of life and death for the people because they trust her. They know she will always do the best thing for everyone. Her position in the tribe is unique because her intuition has been sharpened to such a degree that it is incredibly powerful. She is wisdom itself.

The Crone traditionally was the keeper of wisdom. The ancient matriarchal mystery schools were run by her. She taught the sacred mysteries to all who came to learn. She had creative intelligence that was higher than anyone because of her ability to have both analytical and intuitive ways of thinking working harmoniously together.[30]

The Crone is of the dark. Therefore, she has power and knowledge that go beyond what is normally understood. She has access to realms outside of the logical mind. She has the answers that are needed desperately now.

The Crone and her energy is the one to fix our fucked up society and heal the planet. She is the one element that has been shunned to such a degree. They try to make her impotent, so that the power of authority lies within groups of men. It is obvious it does not work. She is the true solution to our problems. It is wise to reinstate the grandmothers back with the power to lead humans properly. For example, grandmothers would make a perfect police force. Nobody wants to mess with an angry grannie.

## Death

13 represents death. Death is the orgasm of life. Death is when you return to Nature. You become one with the ground. It is the apex of your experience here on Earth, to melt slowly back into her. Death is the secret shadow that follows all living things. People today have it all wrong. Death is a happy occasion. Death is ecstasy. Accept death as the foundation of life. Once you do so, everything else becomes easy.

When you are born and when you die there is a chemical called dimethyltryptamine that is released in the brain. It is a natural psychedelic and creates a relaxed state of euphoria while removing pain. It is Nature's way to make the death process easier to handle.

13 represents death as it is being transformed back into life. It is

the climactic turnaround point where compost turns into dirt. The moment where the carbon of bones turns into a diamond. Where the rotting flesh turns into vulture food. Death equals metamorphosis. The shedding of the skin for a new form to be born. The Law of Perpetual Transmutation of Energy, a Law of Nature, says change is the only constant.

Death cleans the energy of the local in which it occurs. It reenergizes it with the lightning energy of the primordial darkness of the Void. This darkness leaves remarkable sparkle wherever it has been. Death is getting sucked up into the cosmic straw. Nature knows this and looks forward to the rollercoaster of it all.

Like the peculiar smell of the ozone the lightning bolt leaves behind, death naturally purifies everything it comes in contact with. Death creates a tunnel that the spiral rainbow of your soul travels through. It whorls upwards through the dimensions on its journey back to the Void. Your soul is a colourful spark that uses this express tube to go back home.

In the old days, and in some poorer countries now, death is a welcomed part of society. The broken down, dirty, mangy, rough around the edges, and falling apart are still there to be seen. They have not been cleaned and sanitized. They have not been made to disappear. They have not been swept under the carpet. Instead, they are left as a reminder of why it is wise to live life to its fullest extent.

The crooked, mangy cat is the witches' familiar. There is magick in the old, the ancient and the not quite perfect. There is vitality to be found in things that are not shiny and new. The energy of time is a tool to be used. Not taken away because the minority finds it offensive. This stealing away of the well-used is the loss of your power and heritage.

Even though death is a natural part of life, it appears to be the one thing that everyone is most afraid of today. Plastic surgery is rampant. It is obviously based upon trying to escape growing old because of the fear of death. You cannot have a garden flourish, unless you kill the weeds.

Society is in the grip of fear of death. This has created terrible ageism. The world is greedily geared towards the young. It is like this on purpose, to avoid the wisdom of the elders. Wisdom is your inherent ability to stay out of trouble. Especially the kind of trouble the world is in today.

In the days of old, death was worshipped. Then you could enjoy your life because you were not obsessing and worrying over your inevitable date with it. Shamans in the days of old would dig their own grave and spend the night in it. Egyptians had vultures in their birthing rooms. Vultures are carrion eaters and represent the turnaround point between death and life. So their children would never fear death. Instead, they would welcome it with open arms because they were familiar with it.

Nature dies all around you every day, and is reborn the next. It teaches you to accept death as a natural part of life. You die every night when your spirit leaves your body. Yet you are reborn every morning, just like the Sun. You die every time you release parts of yourself that are no longer serving the whole.

Death is present when the spirit person hunts their beliefs, killing the doubting left brain as it protests. Death is lying low when your ego is killed. The constant upheaval of change is something that a wise person welcomes in their lives. It is wise to honour scavengers as the cleaners of life in service to death. This can assist in creating an easy transition for yourself when your time comes.

Death is the art of letting go. There is ecstasy when you do so. The transformation of death always leads to rebirth. Death is expansion. It makes you more, not less. Death and spirit are tightly woven together. A spirit person learns to accept death as their friend. Let it propel you through your initiations that go along with growing and developing your sparkle power.

Death comes as an initiation. It is to be accepted and rejoiced. Not mourned with sadness. Nor the selfish desire to hold on. It is wise to let go of the ones you love, so they may fly high. Death is the soul's return to the ecstasy of the Creatrix. People who are dying are the lucky ones. You feel sad when they pass because you have to stay here alone. You are crying for yourself when someone dies.

Death is the way the Earth takes care of herself. In this way, she maintains the universal balance. It is wise to accept the savagery of death. It takes the old, the sick and the weak. This is how the bloodlines stay strong. It is wise to honour sickness because it takes away the weak so that the strong may have a vibrant quality of life. Honour pain as a wise teacher. She is the servant of death. Honour death, because she knows best.

True power comes from facing your fear of death. A spirit person dies a thousand times over. Each death is a celebrated achievement. It is an initiation and a gateway to a new world. It is to be honoured and revered. It is to be hunted and chased. It is a true blessing. Balance is being immersed in both the dark and the light.

Death is death. There is no "evil" death and "good" death. When the Creatrix wants someone to die, they die.  The Creatrix uses people as tools.  The Will of the Creatrix is done through you.  It is not your place to judge.  All death is natural and it is commonplace.  It comes in many ways and many forms.  All which have their origins in the Will of the Creatrix. "Murder" is a fabricated energetic construct.  It is wise not to do it because it creates heavy karma for yourself that you will have to eventually work out.

Buildings, roads, concrete and metal kill the Earth.  They stunt her energy.  Roads bring death to many of the Earth's creatures.  Can you blame her when she creates an earthquake to take away her numbness in order to reclaim her body, so she can feel that part of herself again?

You feel terrible for the humans who are injured.  What about the damage being done to the Earth in the first place and the resulting pain she feels? The Earth is a sentient being.  She feels everything that is being done to her.  Machines that cause such destruction should not even exist.

You have a date with the Mistress of Death.  It is wise not to be late. The idea when you die is to go up and out.  Unfinished business makes it very difficult to go up and out.  This is how ghosts are born.  It is wise not to fight death, because you probably do not want to become a ghost.

Some people commit suicide in an effort to rush their date.  Maybe you think you can escape what you deem a difficult life.  Most who do may not realize that life in the spirit world is just as tough, probably more challenging, than being alive.  As a spirit who committed suicide you have very heavy energy which makes leaving the Earth planes increasingly difficult.  The end result could be aeons of time as a put-out poltergeist. Something to think about if you are trying to escape reality.  It could get way worse.

death of the ego

Energetic death of the ego makes spirit people. It is wise speak your truth. Take the time to tell people how you feel. This is the wise way of life, to be honest and speak your truth.

You are not responsible for the consequences of your words. You have no idea how someone will react. You are only responsible for being true to yourself by being honest and telling the truth. In return, be prepared to be told the truth about yourself.

By telling people the truth it can assist them in letting go of old and outdated ways of being. Otherwise, they can get stuck. Sometimes people need a little loving push in order to get their juices flowing. Usually such a jolt is nicer and better heard coming from a friend, than a stranger. Though both will do the trick. It is not your responsibility to nudge others, but it can be greatly appreciated when you do.

You are responsible for your sentiment and the intention behind your words. It is wise to make sure you are telling someone how you feel with the intention of their well-being at heart. Otherwise, you can come off like a sharp-tongued, busy body and no one will like you when you are old. You may just create some ugly karma that will have to be worked through at a later date.

Telling people your truth is not about being judgmental. It is about sharing how you feel. It is a fine line. With practice, it becomes easy to be honest. Ultimately, you choose your feelings and then project them outwards. No one can make you feel anything. You choose how to feel.

At the same time, you have feelings based on how the circumstance is rubbing you. This is your intuition. It can be trusted. It is your intuition that you want to listen to, not your judgments. It is wise to learn to separate the two.

Accept the death others give you. Take it seriously by listening to what it is trying to teach you. Then change if you want. Change yourself, your behaviour, and your habits. It is wise to receive death and allow it to flourish inside of you. Honour it as the wise teacher it is. It is a sacred thing, separate from the messenger.

Acknowledge that it wants to break you in order for the new you to emerge. You can surrender to it. Sometimes it is difficult for you to see how you are being unless you are told by the people around you. A real friend will tell you that you are being a gimp.

Say you have a close friend who has been acting like an idiot lately. You could just ignore it. You could stop being their friend. Or you could tell them how you feel.  When you speak words of your truth, it is real, as it comes from the heart.  The person cannot help but hear the validity of your feelings in the moment.

They get a reflection of themselves that they were unable to see.  Then they are willing to do the work to change because they do not like themselves that way either.  The person lets go of that aspect of themselves and they grow into a new person.  This is death.  It is a 360 degree mirror.  There is nowhere to hide.

Telling someone how you feel in a nonjudgmental way can assist people in their evolution.  The sacred mirror of love free of conditions peels away the layers, to get to the core of who you are.  Once the tiers have been lifted away, the true self can shine forth.  True love is about wanting the best for others.  This means telling your truth and enabling their spirit to grow.

Or it can burn the bridge and you will not be friends anymore.  But that is not your responsibility.  It is more important to honour your truth than maintain a false relationship based upon not being honest with each other.  Clean out your closet for new clothes to come.  Before you know it, you will have friends who like you, who can honour the truth you share with them.

"I am going to be polite even though there is a ghoul eating my friend's brains because I do not want to hurt their feelings.  I am sure it will be fine."

When you do not tell people how you feel, you can take on their shit.  This means you are carrying it for them.  This kind of personal sacrifice has been done in the past, but can deeply hurt you and others.  The withholding of information can be detrimental, as it creates shadow.

Speaking your truth is a form of initiation.  It can be uncomfortable.  You can be afraid of pushing the boundaries and limits of others.  For example, saying something to a stranger.  It is wise to keep in mind that you get a reward for doing the work.  You are empowered by doing it.  You are offering them a chance to raise their vibration.  This in turn raises the collective vibration.  And in turn, this raises your own vibration because you are willing and did the work.

This kind of thing creates spiritual atmosphere.  The light of knowledge and awareness dispels the shadows of ignorance.  This atmospheric change benefits the entire Earth.  It does not have to benefit the person.  It does not even have to involve them.  It is for the benefit of the whole.  It does not matter if they respond by making a change or not, as long as you speak your truth.

## spirit death

There are many interpretations made about life that are taken as truth. Where did these ideas come from? Are you sure they are correct? Some of these ideas are inherently flawed concepts purposely woven into spiritual practices to make them less effective.

The following concepts are valid, as they brought you to the place you need to be. But in order to grow, it would be wise to let some certain aspects of the work go in order to develop a new harmonic way of being. It is not that they are wrong. It is just that they are outdated. There are now better ways to do things.

## stonehenge

Take a look at Stonehenge. You go to Stonehenge because you believe it is a place of power. You are correct. It is a place of primordial power. The ancients were incredibly wise, wiser by far than humans are today. They created many megalithic structures that hold power because of the way they are aligned to the Earth, Sun, planets and stars.

So you are excited to go to Stonehenge.  Maybe something mystical will happen to you!  You pay an exorbitant sum of money to be let in.  Then you get one of those little headsets for another ridiculous price so you can listen to someone tell you the history of the place.  You walk around the stone circle enjoying the sights.  After an hour, you leave.  Maybe you are somewhat disappointed you did not have a spiritual experience.

Now it is time to review.  Putting an electrical device on your crown chakra will activate your pineal gland in a detrimental way.  It will shut it down and halt any ability it has to receive higher transmissions and feelings.  It also creates an electrical field around you that dampens your magnetism and interferes with your energetic communications with the beautiful sarsen stones around you.

Did you notice you were walking widdershins around the circle?  This particular Earth vortex will only work if you walk sunwise.  Hey, did you observe that you were walking on rubber mats?  This will completely cut off your energy from your root reaching into the ground.  It will obstruct any vibrations you may have received from the Earth.

Reality is completely controlled.  It does not want you to discover spirit.  It does everything it can to interfere with your process of connecting with your soul.  It wants to meddle in your relationship with Nature.  It knows magick well.  It uses it against you in clever ways.  It is wise to wake up and claim dominion over your experiences and your life.

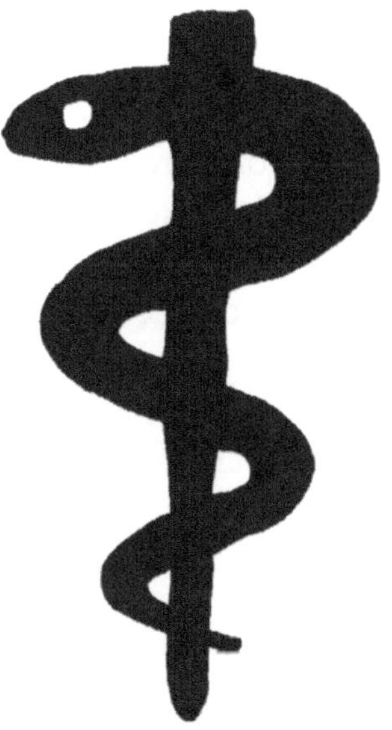

conventional medicine

Western conventional medicine is a panacea because it only treats the symptoms of disease, not the cause. It is an incredibly profitable business that makes pharmaceutical band-aids. As a system, it is not concerned with the true health and welfare of humans. It is truly in service to profit.

Conventional medicine has killed practically as many people as it has saved. Fatal accidents happen concerning the toxicity, and harshness of pharmaceutical drugs. There are many untold side effects created when different drugs are combined. Triage and saving people's lives in emergency situations is one thing. Drug dealing to half-dead people for profit is another.

Vaccines contain a host of poisons, such as mercury. They have not been proven to be safe or effective. It is wise to question everything you put in your body. If you cannot be one hundred percent sure, then it is something to analyze carefully before committing. There are homeopathic alternatives to vaccines that are available and are more trustworthy.

When it is your time to get sick and die, there is a reason it is happening. Doctors do not have the right to interfere with your date with death. They become sorcerous when they try to tinker with your soul. Science is not more intelligent than the Creatrix. It does not have the inherent ability to know the best thing to do. Great damage is being caused to the genetic bloodlines of humans because of this obsession with allowing everything to live.

Energetically, this is causing a backlog with the Mistress of Death, the Dark Goddess, the guardian of the Book of Souls. When you do not meet your maker at the correct time, it causes you and your energetic systems much stress. There are dead people walking around. This is not wise. It has untold stress upon the soul, as it is late for its important date. It is no wonder everyone keeps talking about the zombie apocalypse. Because it is really happening.

So much wealth and resources are being held tight by people who should be dead. Instead of falling into the hands of the next generation as it should and has throughout time. The younger generation would promote the vital changes the Earth needs to return to her righteous self.

Instead, the resources are being hoarded by people who are not quite human because of all the synthetic sludge in their veins that is keeping them alive. People who do not care so much for the Earth, as much as their own survival. This promotion of egotistical self-serving attitudes is killing the planet and all her creatures. It is wise to put the whole first, not yourself.

The Earth is supposed to be managed by human Guardians who are connected to the Earth. Not walking dead people who are not connected to the Earth because the foundation of their life is inorganic. The use of pharmaceuticals is the #1 cause of the fucked-upness of the Earth today as it is the cause of overpopulation. Overpopulation is killing us. Too many resources are being used. The system compensates by fabricating synthetic resources for consumption. This causes sickness and death in humans. All the more profit for them.

Overpopulation has us thinking eating animals is bad. It forces us to worship convenience. It has us believing rampant violence is cool. It has people thinking pharmaceuticals are a-okay and chemical laden, machine made foods are healthy alternatives.

Overpopulation deludes us into thinking plastic is a viable solution. It makes us believe that sickness is normal. It forces us to compromise in ways that are completely detrimental for our well-being. It has the world in the grips of a deadly nightmare.

Overpopulation keeps humans in a pressure cooker. This creates the needed stress that the system profits from. It is no coincidence that over-population is the root of the problem. It is the fuel of the system. Without it they are fucked. The control is so heavy in order to maintain this illusion.

## negative protesting

There is a huge rally focusing on saving the trees. Thousands of angry people yell about the trees dying. How the forests should be saved. This is a classic example of a spell. The energy goes out into the ethers. The trees hear thousands of people yelling about how they are dying. Therefore, the trees say, "Oh shit, I guess we are dying." So they die. What you put out will come back to you. The more energy you give it, the more it grows.

Focusing on the malevolent creates more malevolent energy. Kind of like that Star Trek episode where there is the big evil ball of energy and they try to fire plasma beams into it to destroy it. But it only makes it more powerful. Yep, it is really like that. Now if you had a rally about loving the trees, yelling about how powerful they were, well the trees would feel loved and grow.

Now you have a big rally about breast cancer. You have thousands of women passionately yelling about cancer. Usually they are saying they are fighting against breast cancer. If you fight something, it gets stronger. If you get angry about it, your anger empowers it. Large groups of women yelling and giving angry power to breast cancer only makes it stronger and more of a reality. The result is that more women get breast cancer. With their own power, they make cancer real enough for it to come to the women who are yelling about it. Paradoxical.

What you believe is real. If you believe in the power of cancer, the trees are dying and the corruption having power over you, then that is exactly what will happen for you, especially if you are yelling, screaming and upset.

No wonder the police officer bonks you on the head with his billy club - because you put your head there for him to clout. You are simply completing his anger circuit. If you put your head in a guillotine, odds are you will lose it.

When you give the aggressor someone to aggress against, it completes their cycle of aggression. In other words, the thing that you are protesting wants you to angrily protest because then it can feed off your energy of angst. Protesting is giving them exactly what they want: energy, power, anger and agitation. All fuel for their fire.

Undercover police have been caught on video being the ones inciting violence and anger on purpose during protests because they want it to happen. It is wise to question why they want it to happen unless they are somehow profiting off it. Being arrested and spoiling your entire life does not make the problems go away.

By focusing upon the negative, more is created. That is not to say that these powerful issues do not deserve to have recognition given to them. But what they really need to do is a ceremony.

A ceremony is acknowledging the issue because the truth needs to be spoken. It is calling upon the Creatrix to assist in the moving of the energy. It is people standing in their power and with clear intention stating the desired outcome. In this way, the karma of the conflict can be released.

A ceremony can be people standing in their power, claiming dominion of their opinions. Making it official as it were. Placing themselves not at the mercy of an angry authoritarian. But instead making a public declaration of intent.

Your unconscious does not recognize negative words. "No more war" is interpreted by the brain as "more war." "Unconditional love" is interpreted as "conditional love." "Say no to the dam" is interpreted as "Say to the dam," which in effect is a positive to the brain. "Stop the bad stuff" means empower "the bad stuff." Therefore, when you are protesting you are feeding whatever you are fighting.

The way to truly exorcise something or make it go away is not by telling it to leave or yelling angrily at it. It is by remaining equanimous to it. Turn away from it, maintain your cool and accept it. This is the best medicine for any of the world's traumas. Stand strong and empowered in front of it. It is wise to not lose your cool and go all crazy.

This attitude does not mean you do not care. In fact, it means that you care so much that you are bringing all that caring and concern into your heart. Making it stronger and using that as fuel to become the strongest you can become. So you can be a righteous leader and set a strong example. All the while, using the force of your focus because you know how powerful it is to empower yourself. Instead of focusing on the malevolent, empowering it and then expecting it to leave.

Before you know it, there are many people who too are focusing on empowering themselves. You find yourself drawn together, enjoying a powerful collective. It is based upon self-empowerment through self-knowledge.

Creating, expressing, empowering each other and sharing knowledge are the true ways of freedom. Creating solid movements of powerful people who are focused upon creating enlightened ways of being is a way to create change.

All the ferocious detrimental twaddle in the world is looking to suck you into its psychodrama. As soon as you dance with it, it has you. As soon as it has you, it sucks your vital lifeforce.  By engaging it, you will not make it go away, you will just make it stronger.  Complete the cycle of the aggressor and the energy spins like a top because you are giving it fuel.

Keep your energy and your dignity by ignoring it.  It is alive because somewhere, someone is still using it to learn their lessons.  They have the right.  Just let them be.  It does not mean you have to give it any mind.  Laughter is the best medicine, as the old wives' tale says.

Signing petitions is a great way to empower whatever you are fighting.  Revolution is something that keeps turning over and over again, but never goes anywhere.  No wonder revolutions in the past have never worked.

## zero fat

"Zero fat" can make you sick.  Animal fat is how the human race has survived for billions of years.  Without it, you are fucked.  Your body needs healthy fats in order to make hormones.  It is a vital need.  Without the lubrication of healthy fat your hormones are whacky.  This manifests as moodiness and unstable emotions.

The issue is that there is an abundance of fat available all the time.  Some do not have the ability to control their intake of it.  As hunter/gatherers, humans lived for millions of years by eating a lot of it during the summer to survive through the winter.  Some humans presently are just overeating.

To replace it with chemicals like margarine, hydrogenated oils and modified oils is a crime.  Animal fat is pristine and created by the Creatrix.  Machine-fabricated products are laden with chemicals.  They are toxic.  And you wonder why humans are consumed with cancer.  Fat is important.

The idea is to eat the right kinds and the right amount for you.

Many packages claim to have "zero fat" in an effort to get you to buy them because you think it means they are not fattening. This is a lie. They are fattening because of the sugar. It is the sugar content that makes you fat, not just the fat content.

## exercise

Yoga is not good for everyone. It can overwork the joints. It can strain and tear tendons. It can negatively affect the brain. It can cause muscle, ligament and cartilage damage. It can easily cause major injuries, if not done properly.

Stretching and yoga may release endorphins which feel good in the moment but they are not actually releasing tension or stress. The condition of muscles is held via a program in the brain. This program comes from your emotional state, consciousness, thoughts and unconscious programs. To relieve tension and stress you must alter the original program. A great way to do this is by working with your acupressure points. Acupuncture and different modalities of massage that are based upon the acupressure points work best.

The best way to have a relationship with your body is not to force it, but to work gently with it. You can focus on your posture. You can communicate with your muscles with your mind. Talk to them and feel them. Try to understand why they are the way they are. Usually your repetitive actions are the source of your pain. Change your actions, change the state of your muscles.

Yoga has become very trendy. The traditional art of yoga was based upon working at eradicating the ego. Many modern yoga practitioners have lost touch with this and in fact do the opposite. They push themselves too hard and end up causing more damage than benefit.

Long intervals of cardio is also a modern trend. When cardio is over done is can stress the joints and the body because of too much impact. It can also trigger the famine response in the body because it is so stressful. This can cause the body to shift into lowering metabolism and hoarding fat. Too much cardio also burns fat and muscle at the same rate, and can cause the body to literally eat itself. What the body loves is short intervals of cardio. High intensity interval training is a great way to go. It is all of the fat burning, but reduces the stress. And it is great for the heart.

## racism and discrimination

To love and want to protect your race follows the Laws of Nature. It makes sense that you want to have survival of those who are like you. Black, white, red, yellow, and brown peoples are all totally different. They are visibly different. Ancestrally, culturally, physically and spiritually humans are all different. This is for a specific reason.

There are distinct definitions between the races, to assist humans in being able to reincarnate fully as humans. You need to learn the five distinct and separate lessons that each race offers. As you reincarnate, you are born into a race to learn its lesson. Over lifetimes, you compile all the separate lessons to become a fully functioning immortal human being.

This is how it is today. It does not mean humans cannot evolve to being free of the separation. But for now, we need it. Humans should have the right to want to protect their culture, heritage, ancestors and memories by preserving their distinct races.  You should be proud of your culture.

It is natural for humans to want to keep the races separate by keeping the cultures and the lessons separate.  Humans are all different.  They all have their own specialty traits.  Some forms of racism are the race trying to naturally protect its culture and lessons.

On the other hand, hate is just hate.  Hating other races can happen because of low self esteem or trauma that occurred at too young of an age.  Hate as a response to trauma is inherited and is a sickness.

If there is a problem with the races, it is that they are not in balance.  Some have become more powerful than others.  In turn, some have been disempowered.  This imbalance is based upon the stealing of resources done by a minute elite for selfish profit.

Discrimination happens because some people are more powerful than other people.  This follows the Laws of Nature and your ability to evolve.  It is something you should be proud of.  To discriminate is how your brain functions.  To have rules against discrimination seems like you are stabbing your eyes with little needles just to see if it hurts.  Time and time again, the system lays these false traps of belief and humans unwittingly step into them.

## end of the world

Deer has been walking that path for billions of years, nibbling that bush, listening quietly.  The Earth is billions of years old.  She as an entity is so vast, old and incredibly wise.  Who the fuck are you to say she is dying?

The end of the world is constantly being predicted.  How many times has that deer heard about it?  We think it is all going to end in our time.  They thought the same during the Cold War.  They built many nuclear fall-out shelters to prepare for it.  Surely they thought the same during the Black Death when two-thirds of the population died.  How about  the Great Dying that happened roughly 250 million years ago.  Almost 96% of the creatures of Earth died.  Bet you they really felt that it was the end of the world.

The end that is happening is not to the Earth, but to the system and its suppression of information.  Big difference.  The system wants you to believe the Earth she is dying.  But it is the control placed upon her that is struggling for life, not her.  She is far too smart for that.

The truth about the end of the world is that it already happened.  There was a flood that killed the Earth many thousands of years ago.  The Earth is slowly recovering from that immense catastrophe.  What you need to do is realize that the apocalypse is over.  You already live in the garden of Paradise.

Humans are trained to mimic the beliefs of the system.  The system has put up these fake stucco walls and you believe them.  The system wants you to freak out and lose it.  It wants you to create a world with violence, death and destruction.  That is why it pumps the media full of this apocalyptic mumbo jumbo.  Humans allow themselves to be programmed by outside entities, instead of listening to their hearts.

The system wants you to be angry and violent.  It wants you to be pissed off.  Your angst is its fuel.  Anger and hatred are so easy to choose.  To be aware of what is really going on and then consciously choosing to be accepting and able to express love free of conditions is not so easy.  Having people who choose to be blissful is exactly what the system fears.  It knows that is the way you will escape from its clutches.

But the apocalypse is not what has to happen.  You can manifest your dreams.  It does not have to be your fears.  When you fully realize you are living in the garden of Paradise, it becomes easy to manifest your aspirations.  The trick is to feel worthy.  Maybe people are quick to get on the armageddon train because they feel that is easier than to feel deserving of utopia.

Time to take the blinders off.  Wake up to the fact you made it.  You are already there.  Humans need to wipe the sleep from their eyes, shake their heads and wake up to their true potential and the incredible, luxurious emerald garden of Paradise sparkling all around them.

# organized religion

Organized religion with its rules, structure, formulated beliefs and hierarchical pyramidal network flows against Nature because it is a standardized expression.  Believing in a predetermined set of rules concerning the Creatrix is kind of like fast food instead of growing and making your own dinner.

Humans are individuals who all experience the Creatrix differently. There is no authority above you who can or has the right to tell you what the Creatrix is.  There is not someone who knows the Creatrix "better" than you do.  There is no one who can tell you how to find the Creatrix.

The structure of religion is a hierarchical format designed by the system. How can a fabricated, artificial and linear system create the pathway to a higher power for a spiralized species?  It will automatically build a hard and linear highway when really the ladder to the Creatrix is more like climbing a tree round and round.

As soon as you give your Will to something that you do not and cannot know the prime intent of, then you have given your power away. This is exactly what being a spirit person is not.  It is wise to take what you like from any system.  But also to do and believe what is right for you.  Not what someone is telling you to do.

Religion is based upon a powerful person from the past.  They created a formula for connecting with the divine.  It worked for them. People around them wrote it down in order to mimic it.  This does not mean those traditions from the past will work for everyone.  When you bring them into the present they lose power and become empty gestures.  This does not mean you cannot use these ideas from the past.  They are meant to teach you how to discover your own ways of connecting with the divine.

Most organized religions, unfortunately, are inflexible. They tend to be highly competitive, fighting a great deal amongst themselves. Each puts the other down, trying to be supreme. They even try to kill their opponents. This has been a major cause of war for millenia now. Like any corporation, they want to protect themselves above being honest. Quite hypocritical.

Following rules created by an outside source makes you accept truth without trying to discover it on your own. In a way, this is an expression of giving up. It is wise to never stop searching for your own truth concerning Goddess and God. Everyone's perception of them is different. Everyone has their own way of understanding and expressing their belief.

Everyone has their own path to the Creatrix. No one has the right to tell you how to worship. You have the right to define and worship the Creatrix however it works for you. It is not wise to judge others for their beliefs because this creates karma. We shall be truly be free when the world's people realize all religion comes from the same source. There is nothing to fight about because we are all One.

## coincidence

Take notice when three numbers in a row appear to you. When unusual animals start to show up in your life and bless you with their presence. When names of things are coincidental. When you see the same person multiple times on the same corner. These things are spirit interacting with you. Three numbers in a row is a key. What did you happen to think or say when you saw it?  Take note of it.  It was important.

There is no such thing as coincidence.  Coincidences or synchronicities are pathways leading to a greater truth.  It is wise to look for them.  Honour them as spirit trying to find you.  It is the twist of the 13 communicating with you.  Déjà vu means you are on the right path.  When you spill something, maybe there is a reason why.  It is wise to stop and look around.

Serendipity, karma, fate and destiny are snuggly bedfellows of coincidence.  These are the stage hands of the play that is your life.  Diligently working to manhandle you into the places you need to be, when you need to be there.  Your grand performance is eloquently and perfectly timed.

They are the currents that will carry you where they will.  To struggle against them is useless.  Better to save your strength.  It is wise to flow with them and allow them to carry you.

When you say or do things, they are real.  When it happens, it is happening.  It does not matter if you did not mean to say it, or that you even believe in it.  What is, is.  There are no mistakes.  There are no coincidences.  There are no accidents.  This is how the spirit world functions.  This is how your unconscious functions.

For example, your entire relationship with someone is formulated in the first five minutes of meeting them.  It is wise to pay close attention to what is said.  Suppose they say something casually, like a joke, "I am your worst nightmare."  Do not take it for granted that the words were spoken.  They are exactly true.  Do not be surprised when that person turns out to be just that.[31]

It is wise to pay close attention to what is actually being said at all times.  There is no such thing as a casual comment.  There is no such thing as an offhand joke.  Words spoken are power.  They should be taken seriously.  In this way, you can hear the energetic world communicating to you the true nature of reality.  Coincidences are specifically tailored to communicate with you.  But you need to be open to receiving them. You need to look for them and accept them when you take notice.

Everything happens for a reason.  There is a lesson being presented to you every second of the day.  It is up to you whether you get it or not.  Accidents happen for a reason.  So when something falls in front of you on your way out the door, grab it and bring it.  Who knows?  Everything is connected.

## the rainbow

The rainbow colours of red, orange, yellow, green, are not followed by light blue, dark blue, and violet - but by blue, violet, and pink.  Yay pink! The pink is of an intense vibration and could not be seen when the concept of the rainbow was being formulated by science in the past.  *The Trinity Theory* includes pink as the highest frequency of the rainbow.  It is the violet being suffused with the white light of God, hence pink.  This lack of pink in the traditional rainbow represents the diaphanous nature of spirit.

## old thought / new thought

At one point in history the black swan was thought not to exist.  So now a black swan is said to represent a new thought that has the potential to unravel old thoughts by dis-proving them. At any time a black swan, a new thought could pop up and spoil everything.

When you think of history, it is compiled upon old ways of thinking. The system is made of the accumulation of fossilized thought forms intentionally created to keep humans in a place. A chosen place of consciousness in order that these outdated ideas are turned into fuel cells to feed the hunger of the system. In turn, generating profit for the system.

The system grabbed ideas that were formulated hundreds of years ago by a few select men and these ideas became the law. Oddly enough many of these belonged to the same secret societies. What a coincidence. Unfortunately, for humans many of these ideas are simply outrageously untrue. For hundreds of years society has been building itself upon a wobbly philosophical foundation, which is now collapsing.

Be wary of techniques used to cover up sensitive areas of thought. Like blanket concepts that try to downplay and eradicate import issues. Terms like "conspiracy theory," "controversial," even things that have not been "proven," These are not the droids you are looking for…

For example, human evolution from monkeys has been proven to be a false concept. The idea that humans started out as primitives 10, 000 years ago and have evolved to become what they are today is simply not true.[32] There is evidence of modern humans' skeletons that were found in gold mining caves in Africa 250, 000 years ago.[33]

Science has come to the conclusion that the DNA of humans was interfered with by an off-planet, outside source.[34] It just does not have the courage to tell you. Also, the concept of evolution and natural selection has been proven to be not right.[35]

In the past there has been set up a great conflict between matter and spirit. Spirit has been made more important and more valuable than matter. This is something that is simply not true. Matter and spirit are equal yet opposite expressions of the same thing. Unfortunately, most of these old and broken down ways of thinking have been created by men and their systems of old. This has created an imbalance in the world. As a result, shitty things have, and are, happening. Humans finally have the knowledge and power to forgive and forget the transgressions of those ignorant men from the past in order to move joyfully forward into their vital future.

For millenia now women have been labeled weak, stupid and less worthy than men. At one point the church tried to create the law that women had no souls, yet men did. The feminine ways of thinking and being have been systematically ridiculed and abandoned and worse. But not anymore.

Right now, new ways of thinking are being unshackled and exposed to the light of day. Women and their new and exciting ways of thinking can add and bring balance to the ways of thinking of men. This would empower the Earth and shift it from its present precarious position.

Humans are on the brink of a new way of thinking. One that is based upon truth instead of formulated ideas the system tries to pass off as truth. The truth is tangible, sparkling and present already for you to see, touch and digest. Right now, it is sparkling like diamonds in the Earth. You just have to make the effort to bend over, brush off the dirt in order to see their luminescence.

There are many, many people who are working diligently to uncover the truth for you to easily see. Truth is spreading like wildfire through all the systems of Earth. There is no way to stop it. It means not living in a world based upon lies. It means promoting real instead of fake.

The Internet is a blessing in the way that it makes the truth available to all. Despite the efforts of the system to flood the Internet with lies to make it as confusing for people as possible to understand what is really going on. You see even today that there has been created a genre of writing called "New thought." New thought is the way that humans humans can better get a grip on reality by exploring it through unchallenged and unproven thought forms. They are not wrong. They are fresh and exciting. And no one knows where they will lead.

There is no commitment that needs to be made by thinking in new ways. There is no risk. If old thoughts work, they will be maintained. It just means that there can be freedom granted in the way things are interpreted. Humans have the right to formulate their own opinions and ideas without being governed by tradition. This is the true path to human development.

The world around you is obviously not right, because there should not be so much pain. This pain carried by animals, humans, Nature and the Earth herself will be dissipated by the ability of humans to think in new ways. Thinking in spirals and curves will bring about healing. Allowing the feminine half of your brain to interact in a holistic and balanced way with the masculine side of your brain will elicit change in the world.

Having women and children participating where they have not been allowed before will bring clarity and nurturing to the hurting places on the planet. Allowing children more freedom to grow and learn new things will ease the trauma in society. Tell the children the truth about the world instead of shielding and distracting them from it. They are the ones with the viable answers and solutions to the world's problems. Bringing power to more people will alleviate the constipation of the political environment.

We take for granted how much of our society is dictated by old ways of thinking. It is readily apparent that the old ways are hurting us, the planet and her creatures. It seems that if we do not honestly choose to change our ways, the Earth herself will force us to. It is unfortunate that we seem to wait for catastrophes to be the motivating factor that get us to respond and change.

# paradox

Here is the thing: collectively human consciousness creates. Humans have had overlords for many hundreds of thousands of years who have been in control of creating the system that governs human society. The system puts immense pressure on humans. They easily believe in the agenda the system has created.

The system uses ideas based on dualistic false concepts to manipulate humans. It makes them uncomfortable by using systems of thought that do not work for the benefit of the human. These notions have been impregnated with so much energy by humans who believe in them, that they have taken on a life of their own. The concept is powerless and dead but has been made alive by human belief.

The power of the human is so strong that it can make anything happen, even if it is false. The paradox is, the world as you know it is full of bogus concepts that humans have made alive by the power of their belief. Religion, science, politics and business all run on these false energetic constructs. But they are not real.

Old ideas are magnetic because so much life force has been given to them. They suck people in by creating a magnetic vortex. This can be true with ideas, objects, places and people. It is probably why religion has been successful in the past.

Only in the imaginations of believers do they have life. In true reality, they are just ghosts, illusions, little floaty things that disappear when you ignore them. The whole concept of a fresh paradigm is that humans get to choose what to believe and what to create. Instead of it being dictated and spoon-fed by an artificial, non-human, linear system.

The beliefs the system presents to you seem so real, but they are illusions created by a minority and only powered by the belief of the majority. If the majority believed in magick, there would be Elves and Faeries walking in the grocery stores. That is how powerful our collective thoughts are. And why the system works so hard to maintain its control over them.

Humans have the most powerful tool/weapon in the Universe: their hearts. You are so powerful that you can overpower anything in your way by mere persistence. Simply trust the benevolence of the Universe and your karma. Believe in the grace of the Universe. The divinity of the essence in the moment will overpower anything in its way. Any poison, artificial form of control is rendered impotent by your Will. Your core of vitality will overpower any obstacle.

What you believe is real. If you believe in hell and the apocalypse then maybe you should start to pray. On the other hand, you may believe we live in the garden of Paradise and all you desire shall manifest in the twitch of a hedgehog's whisker. Then maybe you should pack your swimsuit. This is the ultimate expression of life. Believing in yourself. Believing in your beliefs.

The paradox is that there are no absolutes. Even in this book. These words may be bang-on for some. But not everybody sees it the same way. It is only one way to look at things amidst a sea of expressions.

There are so many truths, angles and points of view. It is all how you look at it. Everyone sees things differently. There are so many different ways life can be interpreted. Even conflicting ideas can both be true. It does not seem logical for there to be only one answer. Instead there must be a billion.

Take it or leave it. There are no rules set in stone. There are only ideas to get your juices flowing. Flowing in whichever way they do for you. It is wise to follow the patterns for yourself in order to read their language.

They are presenting to you your gateway to the secret garden.

Paradox is created when balance happens. Balance itself is a paradox because it implies inertia in a world that never stops moving and creating. Often times those peak experiences of balance can be both scary and exciting, perilous and exhilarating. A mixture of pleasure and fear. Apprehension combined with desire.

# The Trinity Wheel of 13

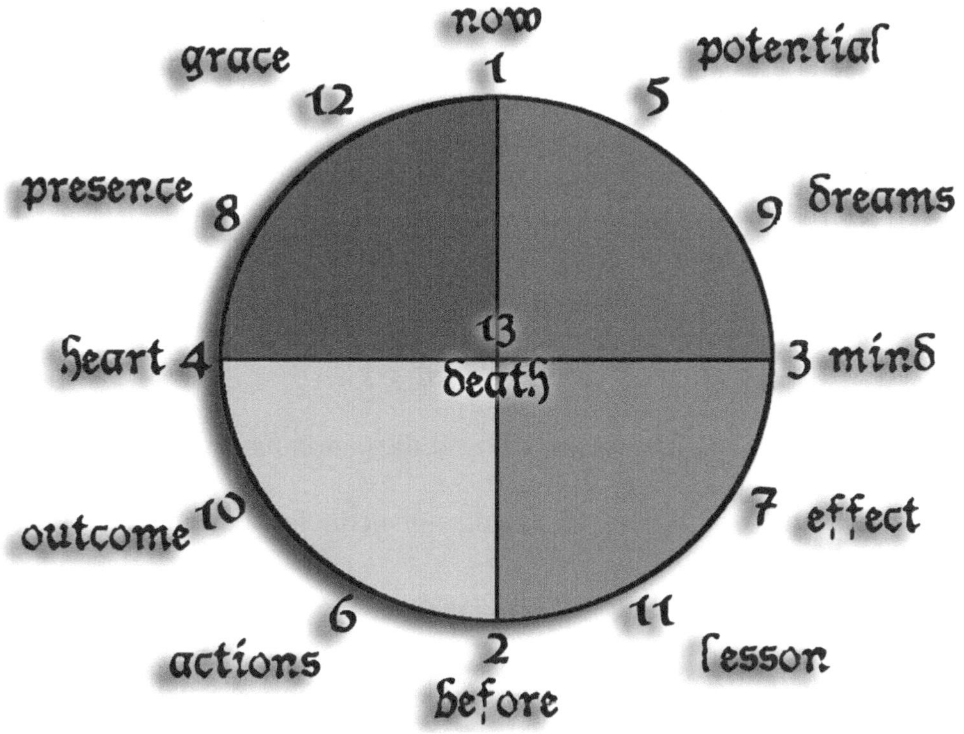

The Trinity Wheel of 13 is based on the number 13. It is a tarot layout that is performed in a spiral, mimicking the spiral of life and thereby following the Laws of Nature. In this way, it works with, instead of against, the flow.[36]

# doing a reading

Draw a Trinity Wheel of 13 on paper, or paint it on wood to be used for readings. Shuffle the deck. Cut into three piles. Pick one. Put it on top of the other two piles. Start with the first card. Lay it in the first position. Interpret the card. Continue pulling cards from the top. A reading is performed in a sunwise direction.

1. Now              ◆ Where you are.

2. Before           ◆ Where you have been.

3. Mind             ◆ What is occurring at the conscious level of life.

4. Heart            ◆ What is occurring at the core level of life.

5. Potential        ◆ Your power to create.

6. Actions          ◆ Your ability to act with your creations.

7. Effect           ◆ The effect your actions will have.

8. Presence          ♦ How others perceive you.

9. Dreams            ♦ Your hopes and fears.

10. Outcome          ♦ The end result.

11. Lesson           ♦ The lesson to be learned for growth to occur.

12. Grace            ♦ For your hard work, there is always a gift.

13. Death            ♦ The mystery card.

After you have laid out the 13 cards, if there are any cards that are unclear, you can always place your hand on the deck and ask, "May I please have more guidance on this card." Cut the deck, and put the new card over the card in question. It should offer some clarity. Repeat as desired.

It is wise to relax and go with it. You can enter into trance. Let a story unfold. You can bounce the meaning of the cards off each other in patterns that appear to you. Maybe the pictures make a connection to each other in your eyes. There are no rules as to how you must interpret the cards. Allow them to trigger reactions. Just flow with the thoughts. You may be surprised at the results.

## Hints for women

♦ You can use ovulation as a guide. If you bleed on the new moon, odds are you will ovulate on the full moon. Women usually ovulate 14 days after the first day of their period. It can be anywhere from 10 to 16 days depending upon the woman. You eventually learn to feel it. You are in total control of your body. You do not have to get pregnant unless you want to. You cannot get pregnant unless both souls will it to happen in the moment of release.

♦ If you do not wish to get pregnant, while having sex, say firmly in your mind, "I Will no baby at this time." Keep in mind, your soul may have different plans for you. The soul always wins over the conscious mind.

♦ It is wise to use a menstrual cup instead of tampons or maxi pads. Then you can keep your moon blood. Your blood is a physical representation of your sparkle power. It can be used in so many ways to improve your life. You can water it into your plants. Use it to mark your grids or boundaries. Mix it into paint. Or dry it into powder to keep it for the future.

♦ Vitamin C is a great morning after pill. It is wise to research it.

♦ The moment of conception calls a soul into the union. The three souls make a contract or an agreement. It depends upon the energy present in that moment as to the vibration of the baby you will create. Superficial vibrational lifetimes call in shallow vibrational souls. You can call a potent vibrational baby to you via your Will. For example, your alchemical state during conception will be crystallized and become the energetic constitution of the child. It is wise to make it as pure as an experience as possible.

♦ The three days of your period are meant to be a gauge of your life. If you are grumpy it is indicating to you what needs to be changed. If you have cravings, it is because you may be overindulging. Your body is pointing out what is out of balance. It is wise to use it as a guide and make some necessary changes.

♦ When you are using the word "she" in a conversation it is called gossip. Talking about others when they are not there can be beneficial because you are trying to learn about yourself, but odds are you are just gossiping.

- It is wise to not work on the first day of your Moon Blood. This is a great day to take it easy. You deserve it. You can do things for yourself. Focus your energy in order to regroup, take stock and figure your shit out.

- Women tend to share so much with each other that it is draining on the energy. Keeping some of it back for yourself is wise.

- It is wise to be proud of your age because it is a sign of your wisdom. The true leaders of Earth are the old women. It is wise to believe in yourself, no matter what the world says.

- Cunt means "the seat of power." It is wise to be a cunt. The world is in a state of disrepair. It needs truthful, confident leaders who will speak clearly and tell it what to do in order to facilitate its healing.

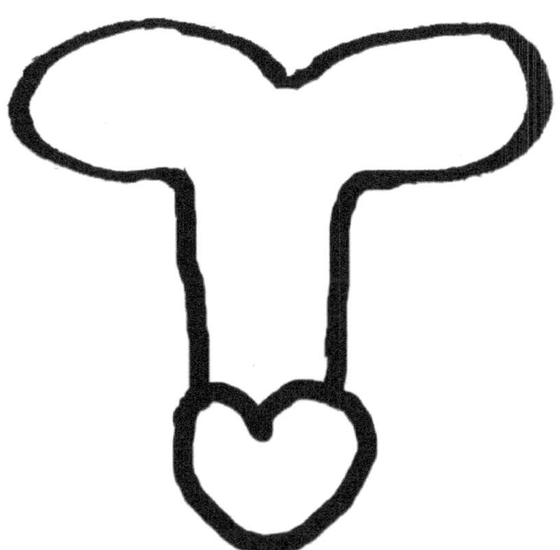

## Hints for men

- Real men cry.

- Men have feelings. Real men are not afraid to share those feelings.

- Men are supposed to take care of the tribe, not kill it.

- Why are so many men aggressive alcoholics?

- If you have a problem, then fix it. If you do not have a problem, then do not fix it.

- It is wise to not make decisions when angry.

- It is smarter to be yourself instead of what you think you should be.

- Realness trumps bullshit everytime.

- You do not have to worry so much.

- It is wise to recognize that there is more to life than ass, sports and beer.

- When dealing with babies and small children, it is wise to relax. Try to put yourself in their shoes. Be gentle, yet firm. Be patient with yourself, eventually you will figure it out.

## Hints for the wise

- It just is.

- Study, learn and research what you can. It is wise to never stop your flow of learning.

♦   It is wise to stretch your mind.  This way it will remain open.  Prop it open with fun.

♦   Taking the hard road offers way better rewards than the easy road.

♦   Do not take things for granted.  Find where they come from, what the ingredients are, what the source is and what it means.

♦   It is wise to have nothing and no one between you and the Creatrix.

♦   Personal rules can limit your life and expression.

♦   The most effective solutions for Earth are spiritual solutions, not physical ones.

♦   Beauty is a mark of the Creatrix.

♦   You are a spiral.  This means you have spiral power.  You can spiral in or spiral out of anything at any time.

♦   You are the creation that needs to be worshipped.

♦   The circle and the cross is the symbol for Earth. The cross was taken out of the circle and put on a stick, so it could be hoisted aloft. This represents the patriarch glorifying his illusory power over the female.

♦   You become whatever you want to become.

♦   Question the intention of others.

♦   Common sense is one of the wisest skills you can grow in your life.

♦   Follow the clichés, the old wives' tales and your mother's and grandmother's advice (odds are they are right).

♦   You have to actually do the work, if you want your life to be of a spiritual vibration.

♦ Spiders have apocalypses all the time.  Imagine how they feel.  They just ride it.

♦ What you put into it, comes right back out.

♦ Everything has a formula.  You just have to figure it out and then it will be smooth.

♦ People are on an elastic.  Go after them and they will keep going the other way.  Walk away from them and they will snap right back to you.

♦ It is wise not to assume things.

♦ Your true power lies within your buying dollars.

♦ Do not give up.  You have to keep at it.

♦ The idea is eventually to be free of everything.  Free of divination, books, gurus and teachings, because you have surpassed it all.

♦ Take out what you put in.

♦ The Sun is your friend.

♦ It is wise not to be a scuttlebutt.

♦ Get lost in wonder.

♦ Questioning everything is wise.  Scrutinize clearly everything you watch, read and hear.

♦ The idea is eventually to be free of everything.  Free of divination, books, gurus and teachings, because you have surpassed it all.

♦ You do not have to trust everyone.

◆ You are alone. You arrived alone via the darkness to this world. So shall you leave. Alone and in the darkness. It is the one thing we all truly share.

◆ All is the Creatrix's Will.

◆ You can know the spiral by farting inside a small closed room. It is like turning yourself inside out. And like the torus, the room spits you out.

◆ Avoid corners. Put a rock, plant or pot in them.

◆ It is wise not to put nails in trees.

◆ Knowing and doing are two different things.

◆ Squirt through the pinhole of perception.

◆ Do not cut down trees.

◆ Follow your heart and your friends will follow.

◆ Seeing a rainbow means you are in the right place at the right time.

◆◆◆

13 represents the mystery of life. Spiritual death is the herald of rebirth. Growth into something new and more exceptional than before.

# ◆◆◆ the trinity theory ◆◆◆

*The Trinity Theory* is a bridge, a rainbow bridge. It hovers and floats right before you, inviting you to come take a look. Just come and step up. *The Trinity Theory* is a sanguine magick bridge to the Creatrix. You have to believe in order to manifest Goddess and God into your life. Other people do not hear and see. They do not see anything at all. But there is more there than meets the eye.

When you have achieved a Creatrix Circuit or perfect balance by bringing awareness to both polarities to make a whole, then something happens. It changes the circumstances. It alters the experience. It brings the Creatrix into the equation. This is *The Trinity Theory*.

When balance is achieved between two, a third thing is created, the rainbow bridge. This rainbow bridge could be anything like soul, a baby, peace, excitement, an orgasm, resolution, a goal, a funny joke, a sacred space, proper digestion, vital health, an exorcism or a smile.

Balance is achieved via awareness. It does not necessarily mean adding two things together or having two things eradicate each other. Simply being aware of your relationship with them creates a certain harmony.

You are a human with a personal rainbow. *The Trinity Theory* offers sincere guidelines for remembering who you are: a well-rounded, conscious immortal human being. Your heart is the most powerful tool/weapon in the Universe. It is wise to use it.

You become like Goddess or God because you are a tiny Goddess or God. A Creatrix Circuit is the ultimate in balance and true harmony. It is a perfect expression of the Laws of Nature. When it is perfect, it clicks.

*The Trinity Theory* is a butterfly net. It pioneers a method of catching an Angel. That Angel you are trying to catch is you. You are an Angel. All you have to do is remember that you are. Your soul is your wings.

In all the swashbuckling between Devil and God, some have forgotten there is another prime character in this play. It is the Goddess. She is the ground upon which the theatre is built. She has very much enjoyed the show, but now is the time for another play to begin. She wants to share with you her way, the way of Sacred Law. She invites you to come to a place within, a place you have never been, a place free of war and sin.

## ♦♦♦ trinity sarah craig ♦♦♦

This has been ~ *The Trinity Theory* Vol. I. The Human Science of Soul. *The Trinity Theory* Vol. II Energetic Guide to Earth and *The Trinity Theory* Vol. III How to Catch an Angel expand these concepts within and without in far directions. They await you... ♦♦♦

You can connect with Trinity Sarah Craig here:
www.trixxcorp.com
www.facebook.com/thetrinitytheory
www.twitter.com/trixxcorp ♦♦♦

You can check out Trinity Sarah Craig's Wheel of the Year calendar here:
www.facebook.com/thetrinitywheeloftheyearcalendar ♦♦♦

## ♦♦♦ notes ♦♦♦

1. Lobsang Rampa, *You Forever* (San Francisco: Red Wheel/Weiser, 1990), 46.

2. Wikipedia, "Triple deity," accessed January 15, 2013, http://en.wikipedia.org/wiki/Triple_deity

3. Wikipedia, "Fibonacci number, " accessed January 15, 2013, http://en.wikipedia.org/wiki/Fibonacci_number

4. Loosely based on the teachings of Chester Grover Ludlow.

5. Wikipedia, "Golden Ratio," accessed January 17, 2013, http://en.wikipedia.org/wiki/Golden_ratio

6. Dan Winter, "Pure Principle based DNA Animation: Why Life Force Whispering in DNA Does Not Want to be Called: 'God or Good or Evil,'" Implosion Group's website about Dan Winter, accessed January 21, 2013, http://www.goldenmean.info/pureprinciple

7. Dan Winter, "The EGGX Files and Bliss Practice - Dan Winter," YouTube video, 1:57:03, posted by "TubeTorusTV," September 22, 2011, accessed March 6, 2013, http://www.youtube.com/watch?v=u1SC0ceLHRU

8. Dan Winter, "The Purpose of DNA," YouTube video, posted by "TubeTorusTV," posted September 22, 2011, accessed February 27, 2013, http://www.youtube.com/watch?v=VPLGN0yIDTg

9. Based upon the teachings of Chester Grover Ludlow.

10. Recommended: *The Secret Life of Plants* by Peter Tompkins & Chris-

topher Bird.

11. Dan Winter, "Star Maps & Bliss Hygiene - Science & Consciousness - The Wind on Which Love Travels, Vol. III," YouTube video, 2:01:46, posted by "TubeTorusTV," September 23, 2011, accessed February 28, 2013, http:// www.youtube.com/watch?v=WJlrnDaeXTE

12. Dan Winter, "Sacred Science of Carrying Memory Through Death," YouTube video, 10:35, posted by "Inphiknit fractal," December 7, 2009, accessed February 20, 2013, http://www.youtube.com/watch?v=U-jntzGTG64E

13. Dan Winter, "Star Maps & Bliss Hygiene."

14. Dan Winter, "The EGGX Files."

15. Eliphas Levi, *Transcendental Magic: Its Doctrine and Ritual* (London: Rider & Co., 1896), 113.

16. *What the Bleep Do We Know*, DVD, directed by Betsy Chasse, Mark Vicente, William Arntz (Portland: Fox Video, 2004).

17. *What the Bleep Do We Know*.

18. Wikipedia, "The Hundredth Monkey Effect," accessed January 28, 2013, http://en.wikipedia.org/wiki/Hundredth_monkey_effect

19. Peter Joseph, Zeitgeist: Moving Forward, YouTube video, posted by "TZMOfficialChannel," January 25, 2011, accessed January 30, 2013, http://www.youtube.com/watch?v=4Z9WVZddH9w

20. Dan Winter, "Redesigning civilization for BIOLOGIC CHARGE COMPRESSION/BLISS-It's Life or Death," Implosion Group's website about Dan Winter, accessed January 30, 2013, http://www.goldenmean.info/lifeordeath/

21. *Magical Egypt*, DVD by John Anthony West, 2005, Top Documentary Films, accessed January 30, 2013, http://topdocumentaryfilms.com/magical-egypt

22. Based upon the teachings of Chester Grover Ludlow.

23. Michael Tsarion, "Architects of Control," YouTube video, posted by "STS2012," June 23, 2012, accessed February 3, 2013, http://www.you-

tube.com/watch?v=vEEqDz_CsQg

24. Frank Lloyd Wright, Interview with Mike Wallace, 1957, YouTube video, posted by "guyjohn59," January 28, 2009, accessed February 4, 2013, http://www.youtube.com/watch?v=nIWkdV38jNU

25. Recommended: Ross Haven, *The Sin Eater's Last Confessions: Lost Traditions of Celtic Shamanism* (Woodbury: Llewellyn Publications, 2008), 140.

26. Dan Winter, "ET Origins of DNA - Science & Consciousness - The Wind on Which Love Travels, Vol. IV," YouTube video, 1:55:34, posted by "Carlo Jonkers," July 16, 2012, accessed February 28, 2013, http://www.youtube.com/watch?v=aWLNyF9H_wg

27. Based upon the teachings of Chester Grover Ludlow.

28. Based upon the teachings of Chester Grover Ludlow.

29. Barbara G. Walker, *The Crone: Woman of Age, Wisdom and Power* (San Francisco: HarperCollins, 1988), 26.

30. Walker, *The Crone*, 65.

31. Gregg Braden, "The 7 Essene Mirrors," YouTube video, posted by "LamatX," May 6, 2012, accessed February 11, 2013, http://www.youtube.com/watch?v=EiBsczafvaA

32. Gregg Braden, "Deep Truth: Igniting the Memory of Our Origin, History, Destiny and Fate," YouTube video, posted by "Bob Loblaw," May 16, 2012, accessed February 2, 2013, http://www.youtube.com/watch?v=wh-KrENfkMEM

33. Drunvalo Melchizedek, *The Ancient Secret of the Flower of Life, Vol. I* (Flagstaff: Light Technology Publishing, 1990), 84.

34. Gregg Braden, Deep Truth.

35. Colin Leslie Dean, "The Refutation: Evolutionary Theory: Natural Selection Shown to Be Wrong," accessed March 21, 2013, http://www.gamahucherpress.yellowgum.com/books/philosophy/Natural_selection.pdf

36. Loosely based upon the teachings of Chester Grover Ludlow.

### ✦✦✦ bibliography ✦✦✦

Braden, Gregg, "The 7 Essene Mirrors," YouTube video, posted by "LamatX," May 6, 2012, accessed February 11, 2013, http://www.youtube.com/watch?v=EiBsczafvaA

Gregg Braden, "Deep Truth: Igniting the Memory of Our Origin, History, Destiny and Fate," YouTube video, posted by "Bob Loblaw," May 16, 2012, accessed February 2, 2013, http://www.youtube.com/watch?v=wh-KrENfkMEM

Dean, Colin Leslie, "The Refutation: Evolutionary Theory: Natural Selection Shown to Be Wrong," accessed March 21, 2013, http://www.gamahucherpress.yellowgum.com/books/philosophy/Natural_selection.pdf

Joseph, Peter, "Zeitgeist: Moving Forward," YouTube video, posted by "TZMOfficialChannel," January 25, 2011, accessed January 30, 2013, http://www.youtube.com/watch?v=4Z9WVZddH9w

Levi, Eliphas, *Transcendental Magic: Its Doctrine and Ritual* (London: Rider & Co., 1896).

*Magical Egypt*, DVD by John Anthony West, 2005, Top Documentary Films, accessed January 30, 2013, http://topdocumentaryfilms.com/magical-egypt

Melchizedek, Drunvalo, *The Ancient Secret of the Flower of Life, Vol. I* (Flagstaff: Light Technology Publishing, 1990).

Rampa, Lobsang, *You Forever* (San Francisco: Red Wheel/Weiser, 1990).

Tsarion, Michael, "Architects of Control," YouTube video, posted by "STS2012," June 23, 2012, accessed February 3, 2013, http://www.youtube.com/watch?v=vEEqDz_CsQg

Walker, Barbara G., *The Crone: Woman of Age, Wisdom and Power* (San Francisco: HarperCollins, 1988).

*What the Bleep Do We Know*, DVD, directed by Betsy Chasse, Mark Vicente, William Arntz (Portland: Fox Video, 2004).

Winter, Dan, Implosion Group's website about Dan Winter - Sacred Geometry & Coherent Emotion, & HeartTuner + BlissTuner, http://www.goldenmean.info

Wright, Frank Lloyd, Interview with Mike Wallace, 1957, YouTube video, posted by "guyjohn59," January 28, 2009, accessed February 4, 2013, http://www.youtube.com/watch?v=nIWkdV38jNU

# ◆◆◆ index ◆◆◆

## ◆◆◆ A ◆◆◆

◆◆◆ 𝓜 ◆◆◆

◆◆◆ ℚ ◆◆◆



◆◆◆ ℛ ◆◆◆

Printed in France by Amazon
Brétigny-sur-Orge, FR